Different V

Images of development
— Colonial
— Rostow
— Brundtland
see p. 17 herein for various terms.
p. 102-03

"development" = subset of "progress"

DIFFERENT WORLDS

Development cooperation beyond the nineties

Thijs de la Court

Translated by Lin Pugh

INTERNATIONAL BOOKS

First published in English by:
International Books,
an imprint of publishing house Jan van Arkel,
Alexander Numankade 17, 3572 KP Utrecht, The Netherlands.

First published in Dutch in 1991 as
Verschillende Werelden, Ontwikkelingssamenwerking in de jaren '90,
by BijEEN, 's Hertogenbosch, The Netherlands.

CIP-GEGEVENS KONINKLIJKE BIBLIOTHEEK, DEN HAAG

La Court, Thijs de

Different worlds / Thijs de la Court. – Utrecht : Van Arkel. – Ill., foto's
Vert. van: Verschillende werelden : ontwikkelingssamenwerking in de jaren negentig. – Utrecht [etc.] : Alliantie voor Duurzame Ontwikkeling [etc.], 1991. – (Bijeenboek ; 76). – Met lit. opg.
ISBN 90-6224-996-5
NUGI 661
Trefw.: ontwikkelingssamenwerking.

Cover illustration by Loek Koopmans
Cover design by OptimaForma
Produced by Uitgeverij Jan van Arkel
Photographs from the BijEEN photo archives
Printed and bound in the Netherlands by Bariet, Ruinen

ISBN 90 6224 996 5

de la Court, Thijs
Different Worlds
Development cooperation beyond the nineties / by Thijs de la Court, [photos: diverse sources] - Utrecht: Jan van Arkel

Contents

Foreword

The relationship between environment and development is very complex. Jan Pronk, one of the leading politicians in the international development arena, until recently Assistant Director of Unctad, and now Minister for Development Cooperation in the Netherlands, has made an analysis of this complex issue. He describes how it was, where it is now, and where it should go in the future. That analysis is his policy paper, *A World of Difference*. It has become the Dutch policy on environment and development.

This book *Different Worlds* summarizes that policy, but because the policy is itself a struggle to get a grasp on reality, to find answers to very pressing questions, this book joins in the discussion.

The context of the book may be Dutch, but it isn't a book about a Dutch government policy. It is about how the West – or, if you prefer, the North – has approached and should approach the issues of environment, development and sustainability. This book brings together environmental, development and peace issues into a theoretical and a practical whole. It also offers a synthesis of ecological and feminist critiques of our world community, giving a basis for the development of community action.

The first four chapters summarize *A World of Difference*. Chapters 5 and 6 deal with the relation between agriculture, the debt problem and the environment. Chapter 7 rounds the book off with a description of the various concepts of development, the perspectives, and the central conflicts in the present development process.

Histories of a number of people from Sri Lanka, Chile and India are spread throughout the book.

A lot has changed since I first started working on this book two years ago. One of the major geopolitical changes in the world took place in the Soviet Union. We can see there, in the changes in the pricing structure, moving toward the "open market", what happens when the individual consumer is not protected from the power of the owners of capital and production. At international meetings, representatives of Eastern European countries identify themselves more and more with the South.

A very positive change is that, with all the discussions within the context of the United Nations Conference on Environment and Development, a new level of understanding has been developed between movements in the North and the South. The "agenda" of the Southern groups is finally being taken seriously by a growing number of Northern NGOs. Political, geographic, economic and cultural boundaries are being crossed. Though the political result of the UN's attempt to do something about the problems of development and the environment is inadequate, at least people are meeting and starting to learn from each other. In the process, many of us have come to the understanding that we cannot change the world without involving the majority of its population, whether they are grass-roots organizations in the South or consumers in the North. People, you and I, are where change starts. We are the ones who have to connect the political frameworks with reality.

Thijs de la Court,
Haarlem, February 1992

Note on the references
In the first four chapters we frequently quote from *A World of Difference.* This is acknowledged in brackets, as "(Pronk, p. x)". The number refers to the page in the official English translation of the policy paper.

Note on terminology
Our definition of "billions" is 10^9.

Introduction

Santiago de Chile, January 1991

While the smog blocks my view of the eroded mountains of the Andes, I listen to Constance. At three-and-a-half she has her whole life in front of her. She chatters away in her rapid Spanish. The future belongs to her, and to her playmates. But what future is awaiting them? Will there still be air to breathe by the time they have children of their own?

In front of me, in the shadow of a tree, are several books which have a lot to say on the subject. One of them has the title *A World of Difference.* Jan Pronk, the Dutch Minister for Development Cooperation, outlines the present situation, takes stock of the past and describes future developments. An impressive work, it would not be misplaced as a basic text for university students of economics. The other books also have depth and strength. But these books have a completely different character. They tell the stories of women, farmers and farming families in Sri Lanka[1], Borneo and Chile. Jan Pronk's texts are global. Those of the other stories are tangible and local.

Jan Pronk's story is important, both for people in the North and in the South. After all, the large development projects, such as the Mahaweli irrigation project in Sri Lanka and the Puyamkuti dam in southern India, change the directions of people's lives. Development projects, thought up, sketched in broad outlines and financed by the North, determine the daily reality of people in the Third World. The farmers of Sri Lanka, Constance living on the outskirts of Santiago, Roaul in São Paulo, Aman in Delhi or Peng in Kuala Lumpur can all tell stories about the differences in this world. Throughout this summary of *A World of Difference* we hear the commentary from the people at the grass-roots level, people living in the dust of the slums or in the mud of the rice fields.

1. DEVELOPMENT: OFF THE TRACKS

Ricardo from Lampa, a small village just outside Santiago, outlines what is wrong. "We have no water, we have no work, we have no houses. If we do have water, it is infested with worms and heavy metals. If we do have work, it is dangerous and heavily underpaid. If we do have houses, either we have to leave because the land does not belong to us, or the houses collapse. We are plagued by rats and drugs, many children are homeless and unhappy. We have no morale, no future and no hope. There is no community, no cooperation, no unity."
Although his energetic and happy face projects a different message, he says: "It is misery, from the beginning to the end, and it is not getting any better."

Ricardo is doing a summer course on ecology and development at the university. "We have to understand what is going wrong, we have to build a community house, give each other lessons, grow vegetables, close the mine and develop alternative employment opportunities. The people can do it, but they have to want it, they have to understand it and most importantly they have to work together!"

1.1 Development: which direction?

"The gap between 'rich' and 'poor' countries has become wider and deeper. It is no longer a question of a relative – and, in principle, surmountable – difference in incomes and national purchasing power. The gap has widened as a result of the decrease in access to the two factors of production: capital and skilled labour. It has deepened because technology increases existing advantages and maintains disparities. The technological revolution in the last few years has left many countries with virtually no prospects: new technology is the preserve of the countries which already have know-how and capital and are able to further develop biotechnology, information theory and other innovations. Technical progress, accompanied by large-scale investment, accelerated means of communication and supranational

> economic decision-making, has made the emergence of a worldwide market possible and changed the world society into *a global village*. But this gigantic village includes a constantly expanding slum area and has so far been unable to make any arrangements for its waifs and strays." (Pronk, p. 35)

The poor get poorer, the rich get richer. The inequality between the two, measured using estimates of the incomes of the richest and the poorest 10% of humanity, has grown sharply. This is reflected in the tremendous inequality in the claims people in the different parts of the world make on the earth's natural resources. The use of materials such as paper and steel is 12 times higher per capita in developed countries than in developing countries. Between 1980 and 1982 an inhabitant of the USA used, per capita, 4.5 times the amount of grain and 60 times as much energy as his or her counterpart in India.

We send people to the moon, look for life on Mars, and become involved in a war which costs billions of dollars. And while the video pictures filmed from a laser-guided rocket show a shot which is perfectly on-target, things are going wrong in the world. Is there something in the mechanism of "development" which isn't working? Because it was the development mechanism which caused the split into two worlds: rich and poor.

> "Hence the lack of prospects: those who are poor stay poor. In many areas of the Third World, and elsewhere, poverty has become hereditary in socio-economic and ecological terms. The process of dualization has a paralysing effect: if little or nothing is to be gained from making an effort, if one's own generation is no better off than the previous generation and if there is no prospect of improvement for the next generation, a dynamic process of development from the bottom up does not even begin. And if it does begin, but the fruits of one's labour are skimmed off – by the elite, by the middle-class consumers in the cities, by large modern enterprises, by other countries – regression or social conflict soon follows." (Pronk, p. 37)

Sunita kneads the dough in a corner of the hut. It is still dark, and her village is engulfed in the smoke of many small fires. The hut fills with sharp smoke from the burning straw in the earthen "chulha", the traditional cooking

> place. Her youngest sister clasps tightly to her mother's breast. Sunita is surprised at the strength with which the baby sucks the last drops from her mother's breast. As thin as a rake, it is impossible for her mother to still the hunger of the baby with only one or one-and-a-half meals a day. Sunita has to stay at home to mind the constantly crying baby, while her brothers are at school. Her mother and father are at work on their neighbour's land. It is harvest time, there is a lot to do. But the work is uncertain. In a few days time, or next week, the work will be finished. Her father can't stand it any more, and has already run off once, because he earns nothing with this work. He thinks of the family as a millstone round his neck. So why did he come back then?
> Sunita imagines herself in ten years' time. As thin as her mother, with children she can't feed. Has it always been this way? Is this a passing phase?

Sunita's life is lived within the almost unscaleable walls of poverty and misery. She will never leave the prison, nor will her children and grandchildren.

The fleeing Kurds from Iraq, the bombed and broken Ethiopians, the inhabitants of Bangladesh who suffer storm surge after storm surge: their misery seems never ending. It neither changes nor even slightly alters the world order. The walls of the prison of the crisis grow higher and higher and its population expands. Slowly, the recognition that things cannot continue this way trickles through to the West. The system threatens to collapse. The crisis of Sunita, of the Kurds, Ethiopians, Bengalis and all the many others stretches out to touch the centres of power. We realise more and more clearly that only a radical change in the world order can stem the misery. This realization forces us to discuss our way of thinking about development itself.

1.2 The sham success of the Western model

> "Policy-makers, in both rich and poor countries, usually regard development as modernization and modernization as westernization. National and local culture was seen as an obstacle to development or as an irrelevant peripheral factor. Various studies have shown how development projects failed because the cultural context was ignored. Where innovations have no relevance to those directly concerned,

> sustained favourable effects are virtually excluded. This finding is important not only for the policies of Western donors but equally for the national development policies of developing countries. In both cases a policy has no chance of succeeding unless all concerned are able to bring influence to bear from the stage at which needs are identified up to and including the time when the policy is formulated, implemented and possibly modified. This calls for an opportunity for a public exchange of views and thus for a process of democratization. Development without democratization will be short-lived; democratization without development is an empty shell." (Pronk, p. 38)

According to Pronk this is the essence of the crisis in development thinking as manifested in the 1980s. The market has become more and more a single entity. Right out into the farthest corners of the world, people are producing for this market. Whether it be the sugar cane in the fields in the Ganges valley near Delhi, the cassava growing on plantations in the mountains of northern Thailand or the coffee grown by Nicaraguan farmers: the international market determines which products are produced and what price is paid for them. With the integration of the market comes the shameless sell-out of the world's natural resources. One country sells its uranium, the other copper, chrome, diamonds or platinum. Many countries sell timber, others sell sugar, coffee or bananas. To be competitive on the international market, the price of these products must be kept low. To keep the price low, no attention is given to the protection of the environment. If enough money is to be earned despite the low price of the products, quantities produced have to be stepped up tremendously. In this way deforestation, desertification and erosion develop.

The Minister for Development Cooperation describes two dilemmas in his policy paper:

> "Is economic growth still possible? The conclusion drawn towards the end of the debate on development theory was that, while it is a precondition for development, it is not enough in itself. Is this still true? Does continuous growth not harm the natural basis for any form of human development? And – the second dilemma – whom does the development process actually benefit? Is there not, in fact, far more at stake than better distribution of the benefits of the process? Is autonomy not the goal? Autonomy for individual people in a society,

so that they have the opportunity from the outset to decide for themselves what development should mean, what life and existence should signify? If that is so, the question of power arises, a choice has to be made after all, development and the achievement of human rights become identical." (Pronk, p. 38)

Sunita's neighbours have started work early. Before school begins the children help with bringing in the harvest. If the weather changes and the harvest has to be brought in more quickly, the children will not go to school at all.

The Banda family have been lucky. Or, to put it better, they have been clever enough to find the weak points in the law. If you're not strong you have to be clever, is Siyathu Banda's motto. They have lived in the new colony of the Mahaweli River Scheme since 1981. The manager of the village thinks of Siyathu Banda as an unfortunate who is not capable of following instructions. He needs extra financial assistance because he cannot keep up with the project. The project leader is very disappointed in him.

Siyathu Banda has created this "unfortunate image" very purposefully. It is the means by which he avoids the obligations attached to the project. The more unfortunate he is, the greater the chance that the bureaucracy will leave him in peace. Besides this he is, as a struggling farmer, more likely to receive extra financial assistance.

Earlier, before the project came into existence, each family had land for growing rice, which was irrigated during the rainy season by water from a small tank. Families produced several varieties of rice. Besides this, they had a form of shifting agriculture. In her family, Mrs Banda was responsible for producing various crops, including legumes, corn, cassava, ginger, sesame seeds, vegetables and capsicums. She was responsible for growing the most important food: millet. The production of millet is typically the task of women. The crops were not grown for the market, but primarily for family consumption. Only the capsicums and the sesame seeds were sold at the local market. Millet was, besides being food, a sort of "savings account" for Mrs Banda. It could be stored for a long period of time, and when money was needed it could be sold, a bit at a time. She used it when a doctor was needed, for the temple, and for the new year's feast.

All the families in the village had livestock: Siyathu Banda had a number of buffaloes to pull the ploughs in the rice fields. The buffaloes were also a kind of savings account for larger expenditure.

Siyathu Banda had lived in the village for twenty years before it was taken over by the Mahaweli Authority in 1981. Everybody and everything had to move. Each family was allotted one hectare of land. As they were not allocated any pasture land, the Banda family had to sell their cattle. They were unable to continue producing food for their own consumption. The use of artificial fertilizer was made compulsory, as was the production of a particular variety of rice. Because the Bandas were no longer able to produce their own food crops, they became totally dependent on the market. They were not able to produce and sell sufficient rice to earn a decent income. Reaping a reasonable harvest was impossible; all sorts of technical things went wrong and the rains fell too early. They had an enormous problem: they did not have the money they needed to buy food. There was little choice: they had to use their agricultural loans to buy food. The situation evolved into a vicious circle. The debts mounted. After a few seasons, Siyathu Banda decided that he had no other choice than to produce his own vegetables, food crops and rice. This was completely illegal, the project managers explained, but Siyathu Banda continued doing things his own way. As punishment, he was allocated less water for irrigation. This did not really affect him, as his method of agriculture required less water than the compulsory method. He illegally sub-let part of his land to friends and relations and rented for his own use a cheap, deserted piece of land which he and his wife used to produce their own food. In the early stages they worked as labourers for richer farmers. Now Sunita's parents, his neighbours, work on his land and he pays them.

1.3 A failing policy

Some countries have to a certain extent been successful in spreading economic and political power over many people and organizations, keeping the most drastic forms of poverty in check. In Western Europe, according to Pronk, three elements play a vital role in this. The first element is the heterogeneous economic order: private enterprise and government complement each other. In a heterogeneous economy there is room for both large and small businesses. Secondly, there are social services. When the unemployed, the elderly and the ill are provided with social benefits, there is a base-limit to poverty. Thirdly there are pluralist and democratic systems of government. People may participate in political decision-making processes, whereby a strength is created which counterbalances the principle of the survival of the fittest. The combination of these three elements has not yet been realized in most African, Asian and Latin American countries.

> "This balance and counterbalance has yet to come to most developing countries. In the international economy, in the global market, it has hardly even been considered. Individual countries reacted to the crisis in the development process with an adjustment policy (adjustment to the norms of the strongest, to the mechanism of the market), while the response at the international level consisted of a few inadequate attempts to make the best of the situation (debt rescheduling for some months, raw material agreements without any effective stabilizing instruments, aid agreements without any real commitment, non-tariff protectionism following the reduction of tariffs, wide-ranging international monetary discussions to put the seal on G7 agreements, etc). They did nothing to check the dualism, to combat poverty or to halt the disturbance of the ecological balance." (Pronk, p. 39)

A failure of policy with regard to environmental protection and the fight against poverty calls for a new approach, which will provide opportunities for growth in all sectors of society. This realization significantly alters the face of development cooperation.

> "Development cooperation's role in this is to create opportunities for people in the Third World so that they may take their destiny in their own hands. No more, but no less either. Nor is this a new role. 'Creating opportunities' has sometimes been called growth, growth designed to increase welfare. At another time it was known as liberation or decolonization. It was sometimes called development to indicate that it was more than just an economic process, that it also had a social and a political dimension. At one stage it was referred to as self-reliance, when there was a better understanding of the conditions that had to prevail if opportunities were to be created for people in the Third World. It has been called participation to indicate that these opportunities had to be seized by people themselves, not for them. Now it is often known as autonomy: autonomy for women, autonomy for people in the Third World in deciding their own destiny, making their own choices as regards the structure of production, the socio-economic order and the political system. Their own choice: their own human rights, their own freedom, their own equality, their own future. Creating opportunities for growth: that is development cooperation." (Pronk, p. 40)

On the outskirts of Santiago, Chile photo: RBP

Outside Santiago lies a vast industrial area. Pastures where cows and horses graze are interspersed with chemical factories, brick-works and homes. Thirty kilometres outside Santiago we take a left turn. A sign saying "Lampa, 23,000 inhabitants" indicates where the hacienda once started. Lampa is the poorest section of metropolitan Santiago and is one of the ten poorest towns in the country. It used to be *one* company, where landless farmers worked for their "master" in exchange for a measly salary and a tiny piece of land.

Land reforms took place in 1964. The landless were given a small piece of land. Most people weren't able to use it effectively. Few people knew anything about agriculture and the seeds were very expensive. Many farmers sold their land and a new pattern of large land ownership was born. Once more the small farmers were wage labourers; 80% of the population are now extremely poor and dependent on work as agricultural labourers. Ricardo invites us into the village house, where we are heartily welcomed.

I listen to a seemingly endless recital of the problems in Lampa. The copper mine, bordering on what used to be the hacienda, uses a lot of water. The waste water contains arsenic and lead which, more and more often, find their way into the village's drinking water. (This is in fact a regularly occurring problem in Chile, where export earnings are dependent to a large extent on copper.) A large part of the land has become salty and is eroded. In the distance, Santiago's industrial area is expanding, getting closer to Lampa. The people in Lampa no longer believe in the productivity of their agricultural labour. They no longer invest in it, and hope for a better future in trade and industry.

We visit Ricardo's home. It is a small house, the size of a garden shed, where he lives with his family – his wife and six daughters. He moved to Santiago twenty years ago, and ended up in Lampa. He repaired television sets. Because technology became more sophisticated, the tools he needed for the repair work became too expensive for him to buy. He now earns a living selling produce grown in his own back yard. His garden is like an oasis in the barren desert. The bees in his greenhouses buzz aggressively in the hot midday sun. Ricardo has taken great pains to bring life back into the soil. Lettuces, cabbages, fruit trees, capsicums and beans are some of the products he grows. Where his garden finishes, the sand begins.

"My neighbours are hard to convince. They have broken their backs, generation after generation, working on this land. Now they hope to work in factories in Santiago. I don't believe this will happen. Even if there was work, it would be the worst paid work, doing the most menial tasks. But it is not easy to convince my neighbours of the value of a healthy alternative. We give lessons in growing vegetables, we have an environmental group, and our village council has joined 25 other village councils in declaring ourselves an 'ecological village'. We research pollution, and involve juridical experts. We have been able to close down a number of companies which were involved in heavily polluting activities."

Ricardo wants to achieve even more than this. But the military camp a little further along, with its watch-towers and the barrels of machine-guns following us as we drive past, reminds us of the years of Pinochet's dictatorship. "That is the burden we still have to bear. We always have to be extremely careful." On our return journey we are talking about forms of concrete cooperation when Ricardo warns me: "Not so fast, not so openly, everything has to be done step by step; the democracy is still so young."

1.4 The system derailed

> "The development problem is not only more deeply rooted than has been realized in the past: it also encompasses interrelated factors whose scope is only now beginning to become clear. Development cooperation is taking on a new face." (Pronk, p. 41)

Poverty is not an accident of the development process. The debt problem is not simply the result of the wrong sorts of policies. Sunita and the Banda family do not live in dire circumstances because of an error in logic and Ricardo in Lampa does not believe that his community was temporarily forgotten and that now, with a modest beginning in a new democracy, the future is open.

No, there are structural errors in our system, in the way we think about development. In the 'nineties, we shall need a very different kind of development cooperation, based on the interrelationship between room to think and room to act.

2. Blurring the borders

"We have to find new land where we have easy access to raw materials and where we can exploit the cheap slaves available in the colonies. The colonies will also be the countries where we can dump the surpluses from our factories."
(Cecil Rhodes, founder of Rhodesia)

2.1 From colonial to neo-colonial

The colonies provided the foundation for Western well-being. Without the gold, silver and slave trade – although in most cases robbery would be a better word than trade – the Industrial Revolution would not have been possible. Europe's rapid growth in production was due partly to the invention of the steam engine, partly to the increasingly efficient use of fossil fuels such as coal and oil, and partly to sales to the colonial markets. Local industries were forbidden, where necessary, so as to safeguard the Western goods from cheap competion.
The colonial era was followed by the neo-colonial era. New relationships were formed, financed by a stream of loans and assistance schemes. The West's surplus of capital and products was diverted to the Third World countries. In this way the billions of oil dollars deposited in Western banks were converted to loans to Third World countries in the 1970s. The loans were spent, more often than not, unproductively. This is the root of the present debt crisis.
The West was not only interested in continuing the relationship of dependence with its erstwhile colonies. "The aid testified to the realization that mistakes had been made which had resulted in a lot of suffering, and as such it could be seen as a form of recompense," Jan Pronk wrote in his policy document.

The East-West conflict greatly influenced North-South relations after the Second World War, particularly as economic bonds with Third World countries confirmed spheres of influence.

"In many respects the Korean war was the catalyst. The frontiers between East and West were rarely fixed in the Third World, unlike in Europe. The conflict in the Third World was far more fluid, with loyalties changing fairly often. Gaining and containing spheres of influence was the name of the game. One way of doing this was by means of such classical foreign policy instruments as the formation of alliances (prompting the Third World to establish the Movement of Non-Aligned Countries in Bandung in 1955, a partly successful defensive reaction) and even the use of armed force. Another approach was to attempt to influence the domestic situation in specific countries, either to reduce the chances of the country defecting to the other camp, or to improve the chances of the country defecting to one's own.

"The aim of such internal activites was to foster political and socio-economic conditions conducive to 'Western' or 'Eastern' development. The socio-economic situation in many Third World countries formed a breeding ground for radical options. The 'East' exploited this in order to bend political developments to its will. The anti-colonial struggles for freedom were a good means of bringing influence to bear, by providing cadre training and military aid. When these colonies gained their independence, they looked on the initially successful Stalinist development model in the communist countries as an example of how their own countries and economies should be run. The East tried to encourage them to take this option by offering various forms of political and economic assistance. It even absorbed them into the 'socialist community of nations', where strategically possible or unavoidable, through membership of Comecon and through mutual military assistance treaties.

"As best it could – efforts were somewhat hampered by the reluctance of some Western countries to grant their colonies independence – the 'West' too did what was necessary to win, or at least not to lose, the battle for the Third World. Economic assistance by means of trade concessions, loans and development aid was an important weapon in this struggle. What had to be prevented, after all, was the rise of social and economic hardship to such a level that a radical left-wing focus became inevitable. A Western strategy of economic and political modernization was presented to the developing countries as the preferable option. More conventional methods of influencing them,

> such as political and military assistance, were also employed." (Pronk, pp. 45-46)

In 1981, Jimmy Carter, then President of the United States, said: "Let us state our position clearly: every attempt by any power to gain control in the Persian Gulf will be seen as an attack on the vital interests of the United States. Any such attack will have to be answered with the necessary measures, including, if necessary, military force."

The preparations eventually resulted in the inevitable: the Gulf War. While the influence of the Soviet Union (occupation of Afghanistan) and Iran in the area provided the first excuse to develop one of the largest integrated army powers in the world, it was Iraq's Saddam Hussein who ignited the fuse. Years of manipulations, in which both East and West selectively invested in weapon supplies and bases, had at some point to culminate in an explosion in the over-armed and totally divided region.
At first it seemed that Iran would tip the scales. This is why the war between Iraq and Iran was so convenient for the West. The arms supplies to Saddam Hussein had a geopolitical character. However, by the time the Iranian threat had declined, Iraq had grown into an untameable monster. CIA support to the Kurds, who are fighting their own struggle in the region, was no longer of any influence. A regional conflict developed to international proportions.

Borders become blurred in many senses when the most advanced jet fighter plane rips the air only metres above a Bedouin village and fires its laser-controlled rockets.

What is at the core of the West's concern in the Gulf region – oil perhaps, or power? Is it irrational fear of the "Islamic peril"? In South Korea and later in Vietnam it was about a united front against the Communist-oriented red peril. When Moscow's powerful leaders transformed into doves of peace and opened their markets to the West, the interest of the West turned to intervention in regional conflicts. El Salvadorians, Nicaraguans and Panamanians have had first-hand experience of this. All these conflicts are really about money and market outlets. Safeguarding the interests of the United Fruit Company in Guatemala, Delmonte in the Philippines and the oil companies in the Gulf are one and the same. The scale of the Gulf War was unprecedented, but that was possible because oil generates so much money, making it financially possible to build a strong military presence.

The Gulf War changed the look of the 'nineties. It confirmed the domination of the United States in world politics. It accelerated the arms race, especially the technological arms race. The world has become even more "one", temporarily. As long as the dependence on oil continues, as long as the export of oil continues to earn gigantic amounts of money for the oil-exporting nations, as long as the many (colonial) conflicts in the region are unresolved, it is only a question of time before the pressure again rises to boiling point.

2.2 East-West and North-South

Eastern Europe and the Soviet Union lost their communist identity. One by one Eastern European countries began to claim greater political independence from Moscow. Other political views have entered the domestic scene, and the way is being opened towards a free market economy. In the Soviet Union the desire for regional autonomy, combined with the ideological and economic crises, crushed the unity and presented a perspective from which state governments could build their own kingdoms. The central government no longer exists, and the various sections of what used to be the Soviet Union are working to build their own economic and political identity. The interdependence between the former Soviet states is very much intact, and for these states to find a common denominator under which they can cooperate in peace seems to be a matter of geopolitical evolution. The economic, social and environmental crises could very well frustrate this process. The enforced introduction of the so-called "free market" confronts the individual consumer with the power of the corrupt and exploitative pricing and trading system which makes it possible for a few to become wealthy through the poverty of many. Neither of the two ideologies, the communist

nor the free market, seems to provide a viable perspective. Whether the Eastern European states will be able to develop their own identity, as a market and as a community, remains to be seen. In this sense, they seem to resemble the "South" more than do some backward provinces of the West.
The influence of the Western economic powers on Eastern Europe and the Soviet Union is now growing incredibly fast.

Pronk is optimistic about the consequences:

> "The struggle for spheres of influence and for influence over individual developing countries, purely and simply to prevent the other party

A young girl training for war in Africa photo: Ackermann/Present

> from gaining an advantage, may become far less pronounced. This is not to say that the developing countries may not remain or become a hunting ground for richer and more powerful states wanting to wield influence, but the East-West motive will recede into the background. This will, for example, reduce the scope for external manipulation of local or regional conflicts. Because of the improved relations between the two superpowers, it will also improve the chances of local or regional conflicts in which both play a background role, or which both fear may escalate into something more than regional conflicts, being resolved or at least better controlled through their mediation.
> "There are also likely to be fewer cases of Third World régimes of dubious moral substance which pursue equally dubious social and economic policies domestically being kept in power or helped to power simply to prevent the country concerned from 'falling over' like the proverbial domino. Financial assistance provided for this reason will wane in importance. It is even likely that the flow of Eastern European aid which, though not extensive, is crucial to developing countries such as Cuba and Vietnam, will largely dry up in the coming decade.
> "The general conclusion to be drawn, therefore, is that Third World countries will be left more to their own devices by the superpowers. The European Community too may be so deeply involved in efforts for the benefit of its own continent in the next few years that little time or energy will be left for the developing countries. Their being thrown back upon their own resources may have positive effects if they see this as an opportunity to develop greater political democracy and economic self-sufficiency. On the other hand, it may make Third World countries into more of a plaything of – possibly less predictable – regional powers. And there is no reason to assume that bloody local conflicts that are difficult to resolve will de triggered by endogenous sources of tension any less frequently than has been the case until recently. Iraq's invasion of Kuwait has recently painfully reminded us of this." (Pronk, p. 49)

Now that the tension between East and West has lessened, there is an enormous demand for financial resources in Eastern Europe and the countries which once formed the Soviet Union. The West is reacting very tentatively to this demand, the result of which is only a very gradual shift in the global cash flow.
In contrast, tensions in the Third World which are by-products of East-West conflicts will ebb rapidly. Soviet and US military support

to Third World countries is being withdrawn, while at the same time peace talks are being initiated. This process will irrevocably lead to more freedom in the traffic in goods and capital in the coming years.

The dust filled Sunita's lungs when the loudly tooting bus drove into the village. Or was it because her cousin Rajiv was on the bus that she was momentarily rendered breathless. Nobody from the city ever visited her, and she hardly knew her cousin. He had escaped the rural life and had actually managed to get a job with the railways. He could read and write. He looked like a teacher, with his heavy glasses, his dark trousers and his white shirt.

Rajiv was a volunteer with an agricultural organization. Because he had a farming background, he wanted to maintain contact with his past. He still toyed with the idea of going back to the country. He hesitated for a moment when he saw how thin his cousin was. Living in the city had developed his political insight. But now he felt that his fortune in being able to read and in having a job had also isolated him. He closed his newspaper, as though he was ashamed of it. Sunita approached him, folded her hands and closed her eyes.

Rajiv read newspapers. He kept up with the world news. It looked as if the superpowers would become good friends. This was the subject of many a discussion in the city. "Now that they have gotten together, the United States will invest in the East, leaving the South with nothing," was the opinion of his friends. "Perhaps that is the case," thought Rajiv, "But would that necessarily be a bad thing for this village? Hadn't Western aid in fact caused many problems? Didn't the people of this village once have their own piece of ground, a small farm, in a well-knit community? There used to be trees, and the women worked together. Now everybody works alone, or as a labourer for someone else. Everything these days has to be so much bigger and faster. Even the very soil is being ruined by chemical agriculture."

Rajiv drank sweet tea at a stall. Dust, smoke and the diesel oil from the buses and trucks distracted him from his thoughts.

2.3 Human rights

The alternative development model offered as an alternative by the former Soviet Union over the past few decades has lost virtually all the support it once had. The viability of the governmental structures

based on this model, as developed in countries like Ethiopia and Angola, is being seriously questioned.
Right-wing authoritarian régimes, at the other end of the scale, are up against the limits of their growth potential. Entire sections of the population are actively involved in protesting against the enormous differences between rich and poor which are cultivated by these régimes. Popular movements which demand democratic reforms are putting authoritarian leaders under an increasing amount of pressure. Because the political, economic and social structures of countries have become more complicated, openly exploitative relations have become less visible. Increasingly, the rules of the game of political decision-making mirror those of the West. The exploitation of people and nature may have gone as far as it can go. In Latin America the military reluctantly returned to its barracks in the 1980s. A wave of political democratization based on the Western model is rolling through Africa, even through South Africa. Political changes are also taking place in Asia. The fact that the Filipinos were able to bring about the collapse of the Marcos régime is writing on the wall. And in the Middle East, the Gulf War has created openings for the introduction of democracy to even the most traditional countries.
Jan Pronk is optimistic. He feels that the time is ripe to look for ways of integrating development, human rights and democracy into one political policy, founded on:
a A pluralistic, representative democracy based on classical human rights;
b A market economy which is ordered, regulated and – where necessary – corrected and stimulated by an actively functioning public authority;
c Public and social services which promote equal opportunities and which provide guarantees against poverty.

What Pronk in effect describes are the foundations of present-day Western European society, which, according to Pronk and many others, should be the model for the entire world.

2.4 The end of the Third World

> "The end of the conflict between East and West, between the First and Second Worlds, also marks the end of the Third World as a separate political identity for developing countries. From now on the trend is likely to be towards a single world system, in which participation in the dynamism of world trade and economic modernization will be more important than the old political dividing lines. The East is seeking to join the West, but in the meantime it has acquired some of the material features of the South. Western assistance can help the East to develop economically and to become politically stable. The South is now in the process of splitting into countries which manage to catch up with the front runners in the world economy, others who are staying the pace with some difficulty and yet others – the largest group – who are in danger of losing touch with the field and falling behind. In almost every country a similar division into population groups is occurring: people whose links with the market are guaranteed by their access to the necessary 'instruments', and stragglers who lack the means (of production) to gain access. One world then, but divided nonetheless, though not along the traditional lines." (Pronk, p. 67)

The phrases "the Third World", "the South", and "developing countries" will continue to be synonymous for a few years to come. The idea that the line dividing rich and poor is the same as the geographic line dividing North and South is outdated. Perhaps it has never been valid. Decades ago we referred to "primitive peoples", unaware of the values and the riches of those societies.

When we want to determine divisions, we have to ask where the power lies. By definition almost, more power means a better social position.

Despite its highly placed position on the scale of well-being, there are groups of poor people in the Netherlands. They are not dying of starvation, many are even healthy and most have a roof over their heads. But they have no say in society, and neither economists nor politicians are interested in them. They have no power. On the other hand, Latin American, Asian and African countries all have rich élites, whose interests parallel those of the powerful in the rich, industrialized countries.

It is difficult to draw clear boundary lines between poor and rich, between powerful and powerless. The terminology will continue, for the present, to be a problem.

2.5 Nuisance value

Development cooperation entails working across the boundaries of national states to create room for human growth and increased well-being for, of and by people. A strong dose of idealism is necessary for this to succeed. It requires selfless participation, not participation for one's own gain. It was this idealism that laid the basis for the rationale of the New International Economic Order which, fed by the strength of the non-Western oil exporting countries, was on the tongue of every Third World organization in the 'seventies. The echo of those times still reverberates today.

Jan Pronk, who once espoused this way of thinking, no longer mentions it. In these times of no-nonsense business-like politics, it has disappeared in the mist of misunderstood idealism. The New International Economic Order is something he associates with day-dreamers, with people who mean well but who, lacking in awareness of what the real world is all about, achieve little.
Idealism is replaced by business-like arguments for international cooperation and the fight against poverty. In a world which is increasingly becoming "one", the positive and negative relations between the various economic powers are extremely important. It is above all these relations – this self-interest in fact – that determine whether the Western world wants to become involved in development cooperation.
Pronk refers to this as "nuisance value":

> "Broadly speaking, there are three phenomena in developing countries which not only pose major problems for these countries themselves but are also leaving their mark on an interdependent world. They are flagging economic development, political instability and ecological disruption. Each has potential nuisance value." (Pronk, p. 65)

Flagging economic development

Millions of Third World citizens experience the effects of the debt crisis daily. Their purchasing power has dropped dramatically as a result of devaluations, life's primary necessities have become virtually unaffordable because their governments have had to put a stop to subsidies, while education and health services are disappearing or have become extremely expensive.

In 1989, Third World countries had to pay US$133 billion to their money lenders because of their debts. This did next to nothing to decrease their debts; the bulk of this sum was payment of interest.

The North also has problems caused by the debt crisis. Because purchasing power in the debtor countries is decreasing, the sales markets for products from the North are shrinking. A striking consequence of the immensity of the debts is the destabilization of the entire banking system itself. Some banks are on the verge of bankruptcy because debtor countries are unable to keep up with their payment contracts.

Indirectly, too, the North sometimes experiences problems following a drop in the world market prices of export products such as coffee and sugar. For reasons of economy, Third World farmers need to move from these to more lucrative products. In Colombia, for example, many farmers have changed over to the production of cocaine. Whether they like it or not, cocaine is their only means of earning an adequate living. Cocaine and its derivatives find their way to the American and European markets through the drugs mafia.

Political instability

The Gulf conflict showed us once again how a regional conflict can grow into an international confrontation. It is just one example of a conflict which is fed by the enormous inequality in the region. The instability of the region is logical: a traditional society, in which an élite has only recently been drawn into the international race of the world market, can hardly be anything but unstable. Where a lot of money is involved (oil dollars, for example), regional conflicts are apt to acquire international proportions. A result of this sort of conflict is a growing stream of political refugees seeking asylum in the North.

Ecological disruption

Through deforestation, air pollution and water pollution, the contribution of the Third World to the problems of the international environment is increasing. The costs of responsible environmental management are too high for the Third World to shoulder alone; the North will have to assist in meeting the costs. The North is in fact party to the disruption of the environment in Africa, Asia and Latin America – take, for example, the polluting factories of Western companies and monocultures which exhaust the earth – and as such should be held responsible.
Ecological disruption leads to flagging economic development in the Third World. This affects the interests of the North. Now that refugees from the regions ruined by drought and deforestation have reached the North, it is actually in the interest of the North to combat the ecological crisis in the Third World.

The days of idealism in our society have gone, it seems. We apparently need to know that it will be to our own betterment before we are prepared to become involved in development aid.
Jan Pronk compares the situation with a racing circuit. The Western countries are numerous laps ahead of the Third World. The problem is that the Western countries have to go slightly off course to pass competitors whose pace is slowing. That costs time and involves taking risks. There are two inconsistencies in Pronk's comparison. Pronk overlooks the nuisance value the North has for the South – which is immeasurably larger than the other way around. Secondly, Pronk describes a world in which the North and the South are racing on the same track, the course determined by Western norms and values.

> Rajiv didn't last long in Sunita's village. The life of the poor farming families seemed hopeless to him. He was in a continual state of conflict with himself. On the one hand the farming life appealed to him, but he didn't, on the other hand, care for the compulsory cultivation of rice. It was an illusion for the small farmers to think that they could establish the price of rice. That was in reality the prerogative of the middlemen and the large landowners.
> Whenever a farmer was no longer able to meet his payments, a large

landowner was quick to respond by paying him a visit. An extra piece of ground for the cultivation of rice was always a lucrative proposition for wealthy landowners. They would simply kick the poor farming family off their land.

Rajiv observed how the Banda family, Sunita's neighbours, had outwitted the agriculture specialists. The Bandas used traditional methods to grow their own food on a small, out-of-the-way piece of land. But if the price of rice rises, the Bandas will also suffer. The wealthy landowners will want to buy this patch of land too.

Nuisance value of the North

The North has been able to develop because it has exploited the South. Slaves, gold and silver, spices, tropical wood and all the other products from the colonies form the foundation on which Western development has been built.

As a result, we now have a situation in which 20% of the world's population has 80% of the world's raw materials at its disposal; a situation distinguished by flagging economic development for millions of people, political instability and far-reaching ecological disruption. In turn, the North is inconvenienced by the effect this has on the "rich", and is forced periodically to deal with the worst of the excesses by waiving a portion of the debt, if that debt is a "nuisance" to the North. (By waiving the debts of the poorest countries because these debts affect the credit of foreign companies, industrialists from the North are once more at liberty to invest in the debtor countries.) This form of aid is not development cooperation. It is a form of economic self-interest, a compensation for the nuisance value Northern countries impose on Southern development prospects.

The extraordinary thing about a term like "nuisance value" is that it avoids the essence of the crisis. It suggests that anybody could take the lead, which in reality is not true. Poverty is not a question of temporarily lagging behind: the phenomonen is inherent in the system. The "leaders in the race" are only able to run in the lead because they have the power to exploit the "stragglers".

As long as the power base itself is not up for discussion, the situation will remain as it is: hopeless. Poverty and the environmental crisis are

Traveling in Bangued, the Philippines photo: Herzog/Present the results of a conflict in which a small group claims ownership over the resources of a large group. To talk about the nuisance value which the large group of people living in the margins has on the privileged few avoids the crux of the problem.

2.6 The race

Pronk's model is built on the premise that the South has to participate in the race. He even refers to "stragglers" and "hangers-on":

> "There are signs of a division, partly splitting the South, the East and, to a lesser extent, the West, into 'front runners', 'hangers-on' and 'stragglers'. The front runners are the countries whose economies feature a high level of technical innovation in both products and production processes and in refined methods of organising production, distribution and marketing. The hangers-on are the countries whose economies will have enough comparative advantages in some sectors to keep up with the front runners to some extent and even to derive some benefit from their activities. The stragglers are the countries whose economies do not have sufficient vitality of their own to stay the pace." (Pronk, p. 61)

According to Pronk, the North will have to help the South move in the same direction. But the practice is that the model which is enforced by the North does not work and that the North is experiencing the negative effects (debt crisis, ecological destruction and political instability) as a nuisance. According to this way of thinking, the process should be streamlined, getting all participants running in the same direction once more. Streamlining can involve trade agreements, debt rescheduling and development assistance.
What would happen if a country no longer wanted to participate in the race? It could, for example, decide to focus production on the needs of the domestic market. It could partially close its borders to Western imports. The North would be furious and would stop all development aid immediately. It would use all methods possible to regain access to that economy, the source of its profits. If necessary, the North will use covert methods, and weapons supplies can also be used to achieve its goal.

Some poor countries try to resist this pressure, but in the end they all have to play the game according to the Western rules. Pronk says the same thing:

> "In the emerging world it will no longer be a question of protecting the developing countries against the weight and conflicts of the developed world, but of promoting or enabling their integration into world policy and the world economy." (Pronk, p. 69)

An African parable

A black man is lying asleep under the shade of a giant mango tree. A white man walks past, sees him lying there and wakes him up.
"Get up, sleepy-head, you're wasting your life away. Why don't you climb in the mango tree and pick the mangoes?"
"Why should I pick the mangoes, white man? The mango tree drops its fruit every day. I collect them for me and my family. The rest will only go to rot in my shed."
"You shouldn't leave them to rot in your shed. You have to take them to the market and sell them for money."
"I see, white man: I can save the money, that won't rot."
"No, silly, you shouldn't save it. Money should work."
"Fine, white man, I'll throw a big party for all the people in my village. And you can come too."
"No, no, no, no parties! That's throwing money away. You must spend it wisely."
"How, white man?"
"Listen. From the money you earn at the market, you have to buy first one, then two and finally all the mango trees in this village square."
"But those mango trees don't belong to anybody. Everybody can eat that fruit."
"That is the whole point. Now they're everybody's. If you buy them, they'll be yours. No-one else will be allowed to collect that fruit. If anyone wants to eat a mango, they'll have to buy it from you at the market. I can see it all now, you'll be a rich man. And then ..."
"Yes, white man?"
"Then you keep on going, buying more and more mango trees in the region. One day you'll own all the mango trees in the country."
"Why should I want to own all the mango trees in the country, white man? The fruit will cover our village in an enormous rotting mass, because

> nobody can eat that many mangoes."
> "Quiet, listen. Of course your village won't want to eat all the mangoes. Conserve the mangoes and build a cannery, where you process and can mangoes. Export the canned mangoes to the rich countries in the North, where very few people know how delicious mangoes are. Your little factory will grow into an enormous factory. Many of your compatriots will work in your industry. And you, you won't have to work any more!"
> "And then what do I do, white man?"
> "You'll have built a large house in this place, where it all began. The garden will be what used to be the village square. You'll string up a hammock between two mango trees, and you can rest in their shade."
> "But I'm resting in their shade now, white man."
>
> (From: Parabels van Grote Spin, Bob Papelard. Publisher: BijEEN)

2.7 The world village

> "The disappearance of the First, Second and Third Worlds provides a unique opportunity to unify the world, to create one world instead of three or even more ... This one world system will be heavily biased towards Western institutions and values. The changes of the 1980s have proven them stronger and more attractive." (Pronk, pp 68-69)

Is this now what the Banda family is waiting for? Does this unification make Ricardo happy? Will little Constance from Santiago in Chile have a sunnier future in a world built on Western values and institutions? The unification seems irreversible. Like a whirling eddy, the North slurps everything up, destroying the identity of thousands of people. Pronk warns against triumphalism, even questions it. But he sees no other way.
"The cultural diversity of the world which we knew two hundred years ago, before we created this multinational culture which is more dangerous than a multinational company, has gone," says Anil Agarwal. The world-spanning might of the big financial powers has an integrating effect. They draw in every aspect of the world to their market system, orienting the dynamics of the varied local and regional socio-economic systems to align them almost magnetically to the dominant market system. Anil Agarwal, along with many other

Third World people, describes this as one of the foremost causes of the present crisis.

The opposite of integration is de-linking. De-linking involves respecting the particular character of different cultures and states – preserving languages, economic styles, religions, cultures and identities. Pronk uses the word "autonomy" to describe this. De-linking does not mean stimulating total isolation, economically and culturally, but stimulating a form of development whereby the different peoples of the world can determine their own path to development in relation to the other peoples.

3. GREATER RISKS

"Anyone who believes that exponential growth can go on forever in a finite world is either a madman or an economist." (Kenneth Boulding)
"The continued pursuit of economic growth by Western societies is more likely on balance to reduce rather than increase social welfare."
(Ezra Mishan)

> "At the heart of the problem of continued ecological destruction is the way the expanding, open economic system functions in a finite, closed ecosystem." (Pronk, p. 78)

The earth is the closed space in which we must live, together with all the plants and animals. The problem is that human activities are increasing so much in volume that the space allotted to us is being used up. Our playground on earth is becoming so small that it is becoming impossible to avert a crisis.

> The tenant of a house walks somewhat hastily to his landlady. He sees a sign: smoke. It is warm, and in the middle of the smoke, red and yellow colours flicker and dance. What could it be? He and the landlady take out books to study the phenomenon, and dial the number of the town hall. The town hall initiates a broad study, and concludes that it is, most likely, a fire! Meanwhile, the landlady is beginning to feel uneasy. The phenomenon expands, and is consuming the entire top floor. The roof is now smouldering. The fire brigade deliberates on which sort of material it will use. Telephone calls go back and forth on the colour of the flames, the thickness of the smoke and the velocity of the wind around the building, but the information is inadequate. The Fire Chief comes to take a look at the situation for himself. He pushes the emergency alarm, and calls neighbouring villages to bring their fire hoses. But the neighbouring councils need more information before they can make a decision. They come to take a look and to discuss the problem. Long discussions take place on liability, insurance and the risks involved in extinguishing the fire. The tenant and the

landlady save what they can, clothes, furniture, books, whatever they can carry. Very little is left when the house is finally engulfed in flames.

3.1 Environmental capacity and economic growth

Before discussing the environment and our influence on the environment, it is important to agree on the terms we use. For a long time we have talked about the "carrying capacity" of the environment, by which we mean the capacity of the environment to bear a certain pressure, such as the pressure brought about by human activity. For example, if we allow too many sheep to graze on an area of land, that land will erode and become debilitated. If the soil quality is poor and

Systemmatic removal of forest in Sindhuli, Nepal photo: Lemmens/HH

provides little food for the sheep, the process of erosion will be faster. The term "carrying capacity" unfortunately does not take into account the relationship that we, people, have with our natural environment. That is why Jan Pronk has adopted the term coined by the environmental economist Hans Opschoor, "environmental capacity", in his analysis of this complex subject matter.

> "There is a strong interaction between the environment and economies. All the physical inputs to the production processes on which economic welfare is based are supplied by nature. During and after production and consumption all these inputs are sooner or later returned to the environment in the form of waste substances.
> "In essence, nature performs three vital economic functions. First, the natural *production function*: it is a supplier of food, primary products and materials, which include both renewable and non-renewable resources. Renewable resources are those produced by ecosystems, either manipulated by humans or not; non-renewable resources are minerals. Secondly, the *assimilation* or *regulation function*: the environment's regulating and stabilizing capacity, including its capacity to process wastes. Thirdly, nature has an *information function*. It provides species of plants and animals which are used as crops, raw materials or livestock. It contains genetic information which is necessary for the renewal and viability of domesticated varieties and breeds.
> "All the economic functions of the environment can be summed up in the term *environmental capacity*." (Pronk, p. 79)

We find these three functions in tropical rain forests:
The production function. People living in the forest live off its products. They have done so for centuries, without damaging the forest. Shifting agriculture as practised by the Dayaks, one of the peoples of Sarawak in Malaysia, is practised so ingeniously that almost none of the fertile top layer of the soil is lost. Erosion seldom occurs in many other traditional forms of agriculture. The emphasis in all these agricultural forms has always been on "renewable", that is sustainable, land use.
An example of an unsustainable use of the production function of the rain forest is the export of tropical hardwood. The damage caused by logging for the export of tropical hardwood can be compared to the damage done by mining. You remove something, without returning anything. At one point or another production, and the land, will be exhausted.
The term "mining" is used for many forms of unsustainable land use. The

highly productive and export-oriented cultivation of coffee, cotton and peanuts are examples of mining.

The regulation function. A rain forest is like a sponge. The fall of the tremendously heavy rains is broken by a thick layer of foliage. In the depths under the foliage, mosses suck up the last droplets of water. The rain forest very gradually releases this water, so that the rivers can stay reasonably full even in the dry season. This regulation function keeps the climate constant.

The information function. We learn a lot from the forest. The Northerners learnt which products were successful as export products. We owe many of our foodstuffs and medicines to our knowledge of the rain forest. The varieties we presently cultivate, however, will die out if they are not regularly cross-pollinated with wild varieties. The rain forest shelters thousands of unknown varieties of plants and animals which could possibly be important to us. We are destroying them now, without even giving ourselves the chance to find out what they are. The most important information the rain forest communicates to us is how an extremely complex system of plants and animals can sustain itself.

Economic growth

If we want to understand the effect of our economic actions on the environment (the environmental capacity), we first have to find the words to describe those actions. The most commonly used term is "National Product" (the Gross National Product: GNP). Economic growth is mostly expressed in the growth of the national product per head of the population.

Pronk is not happy with the present definition of the concept National Product.

> "There are growing signs that the pre-eminent indicator of economic success – GNP – does not correlate well with actual economic welfare. It does not, for example, say anything about its distribution, and yet fair distribution is an important component of sustainable development. Nor does GNP in itself take any account of the reduction of the environmental capacity. A better insight into national income can be obtained if reductions in the value of natural capital and defensive expenditure for the environment are also considered. Such expenditure is undertaken to prevent or combat undesirable side-effects of economic activities on the environment. It includes, for example, the cost of water purification, some health care and measures to counter

erosion and desertification. Defensive expenditure now forms part of the national product, when it should essentially be regarded as an intermediate cost item and thus omitted from an indicator of changes in the level of welfare." (Pronk, p. 80)

To show where present calculations of GNP go wrong, researchers from the World Resources Institute have made some alternative calculations of Indonesia's GNP. They show that, given the country's shrinking oil reserves, soil erosion and deforestation, the economic growth of Indonesia is not 7% but 4%.
Only a limited number of factors have been used in this calculation. The fact that the oil has to be transported, burnt or converted has not been taken into consideration. Pollution results from these activities, which perhaps does not affect Indonesia. To what extent is the pollution of the seas taken into account? Or the acidification in Western Europe, caused by this oil? Could economic growth, as conventionally calculated, in reality be a recession? Here we are, borrowing from future generations, our children and grandchildren, while knowing full well that we cannot repay them ...
These sorts of informative calculations can be made for any country. If in Zimbabwe, for example, the loss through erosion of the two essential soil nutrients, nitrogen and phosphorus, were to be compensated for with artificial fertilizers, that would cost US$2.5 billion per year, equivalent to half the annual GNP for that country.

Table 1. United States: Growth of GNP and Index of Sustainable Economic Welfare (ISEW) 1951 - 1986

Period	*Growth GNP/person*	*ISEW/person*
'51-'60	0.97%	0.84%
'60-'70	2.64%	2.01%
'70-'80	2.04%	0.14%
'80-'86	1.84%	-1.26%

source: Daly H.E. *et al*, *For the Common Good*, Boston, 1990

Pronk quotes a study on the United States economy. Researcher Daly and his colleagues calculated an indicator for economic welfare, which takes into account the depreciation in value of the natural capital (forests, fertile land), the defensive costs (cleaning, combatting erosion), and also elements of unequal income distribution, and costs of urbanization and unpaid labour. This index for sustainable economic welfare shows that the economic welfare per person in the United States has been declining since the beginning of the 'seventies. The results of this study are reproduced in Table 1.

> Herman Daly and John Cobb, two prominent advocates of a "steady state economy" (an economy in which material growth is not the ultimate priority) show in their book *For the Common Good* that welfare is a very relative concept. They argue that the feeling of personal satisfaction is largely dependent on the position a person has in relation to others; in other words, it depends on the relative status of the individual in a community. Generally people with a high social status indicate that they feel happier. That is to be expected. But when comparisons are made over time or between different human communities, it appears that the degree of experience of happiness differs very little. The fact that at a later date most people are consuming more in absolute terms does not have much effect on happiness. People in prosperous communities are in general no happier than people in poor communities.
>
> In other words, the absolute level of the economy is no measure of personal satisfaction, while the relative economic level within the community is of very great influence.
>
> The disadvantage of the term GNP is that it offers no information whatsoever on inequalities in a society.

You don't see reality, you interpret it. It is not possible to understand the world, but the centuries of money economy have made it seem simple.

> If you take the motorway from London to the north, you find yourself jammed in between a stream of lorries transporting biscuits from London to Glasgow. And if you look at the traffic travelling in the other direction, you will see just as many lorries transporting biscuits from Glasgow to London. An unsuspecting observer could come to the unavoidable conclusion that biscuits have to travel at least nine hundred kilometres before they are fit for consumption. This was an observation made by E.F. Schumacher, noted in his book *Small is Beautiful*, bearing on the "scale" of production. The gigantic biscuit factories in London and Glasgow could just as well make extra cookies and sell them elsewhere. "Of course", says the biscuit maker in London, "I sell as many as I can locally. But with that quantity my ovens

> aren't filled. And because I have to keep the ovens alight for biscuits for London and the surrounding area, it doesn't hurt to add a few biscuits for Glasgow as well. That doesn't affect the costs." But our baker demands the same price for his product in London and in Glasgow. His excess production is almost pure profit. He is not interested in the fact that his profit will largely be consumed by the cost of transport to Glasgow. The biscuit maker in Glasgow thinks exactly the same way. He too wants to use his factory to its maximum capacity. He too will try to conquer the London market with his excess biscuits. Looking at the two situations side by side, what seemed logical and rational from the point of view of the biscuit makers is a complete farce. That is the fascinating thing about the economy: there is a logic in relation to individual companies which is ludicrous if you look at the overall situation.
>
> And so we send our products to all parts of the world, wherever there is purchasing power. We use cheap energy, in the form of oil, gas or coal, to make the transport of these products possible.

People assume that money can buy anything. Many economists claim that North Africa, for example, should export more products in order to attract strong foreign currency. With this foreign currency, North African countries can invest in their own productivity (once more for export), repay debts and finance essential imports. If hunger is rife and the soil is infertile, it is not a good economic proposition to invest heavily in agriculture for the domestic market. The inhabitants have no money, so they cannot afford the necessary investment in artificial fertilizers, pesticides and irrigation. According to this way of thinking, it is more lucrative to invest in modern forms of export agriculture. With foreign currency, grain and other food for the country's starving population can then be bought on the world market. Through technological improvements and higher efficiency, made possible by temporarily improved access to the market, these hunger-stricken countries are able to step out of the poverty spiral.

That's why we eat beans, capsicums and other products from Africa, all of which have been produced by a starving population.

The theory of these economists make sense, as long as you assume that anything is possible if you have money. But production for export is a form of mining. It requires an intensive use of energy and fossil fuels – for the manufacture of fertilizers and pesticides as well as for transport to the West. To grow crops for export, the best land

and the best soil is used. The poor population is left with the fields which yield less. To be able to produce anything at all, the poor have to use the land very intensively, and this inevitably leads to soil erosion and desertification.

Sustainable economic growth

> "The ultimate aim of policy at the national and international level will be sustainable development as defined by the Brundtland Commission: 'Sustainable development is development that meets the needs of the present without compromising the ability of future generations to meet their own needs.'
> "Like most other definitions of sustainable development, the concept comprises in effect three components: growth of production (economic development); fair distribution (fight against poverty); and the maintenance of environmental capacity (ecological sustainability).
> "Where the sustainability component is concerned, the Brundtland definition needs to be further differentiated to reflect the limited substitutability of natural and produced capital. In the case of *weak* sustainability, the two forms of capital are regarded as complementary and totally substitutable, and an attempt is made to maintain or increase the sum of the two. In the case of *strong* sustainability, the supplementarity of the two forms of capital is recognized, and an attempt is made to maintain natural and produced capital separately. In general, policy will have strong sustainability as its ultimate goal."
> (Pronk, p. 82)

It is not easy to pin a price tag on the environment. There are two reasons for this. First, "produced capital" is dependent on natural capital. Nature is necessary for production. If we were to convert all natural capital into money, we would ruin our economy. In other words, our economy can only function on the condition that natural capital is conserved. Secondly, natural capital fulfils a large number of economic functions: climate regulation, protection of water supplies and maintenance of diversity. These functions are difficult to express in monetary terms. Nature is multifunctional and can be used at many different times for many different things.
Nevertheless it can sometimes be useful to demand financial compensation for degrading the environment, above all because it makes

more environmentally benign activities demonstrably cheaper. The aim is to develop price tags with which decision making can be influenced in the direction of environmental compatibility. The price tags are not meant to be a means by which a balanced "budget" is developed, in which the "value" of the environment has been calculated in real terms.

Unfortunately, economic activity in the world has, to date, not been determined by this sort of price comparison.

3.2 The population explosion

At present there are 5.3 billion people in the world, of which 4 billion live in the Third World. Ninety million people are added to the world's population each year, and most of these people are born in the Third World. Extrapolating this trend, the world will have an estimated 10 billion inhabitants at some point in the next century. All these people will need to eat, consume products, throw away refuse and be mobile. Can the earth stand up to demands of these proportions?
It would be wrong to say that population growth in the Third World is responsible for the deterioration of the environment. The average Dutch person consumes per year, for example, the same amount of wood for all his or her paperwork as an average Indian does for cooking and heating.

> "Although the rich countries account for only 5% of the total annual growth of the world population, this increase makes as heavy a demand on the environment as the far higher growth in developing countries." (Pronk, p. 87)

The Brundtland Commission wrote in 1987:
"By the turn of the century, almost half the world will live in urban areas – from small towns to huge mega-cities. In the 21st century the world will for the greater part be urbanized. Over only 65 years, the developing world's urban population increased ten-fold, from around 100 million in 1920 to close to one billion now. In 1940, only about one person in 100 lived in a

city with a million or more inhabitants. By 1980, one in 10 people was a 'million city' resident."[2]

Tjeerd Deelstra *et al* list the following as environmental problems caused by urban living[3].
"Several eye-catching examples of environmental problems in cities are:
- loss of fertile soil within and near the urban area – this is appropriated by the building industry;
- deforestation near cities, causing erosion – this results in periodical dust storms, making the cities less livable, and in earth movements, destroying residential areas;
- pollution of soil, surface water and ground water, resulting in a shortage of pure drinking water, an improperly functioning drainage system and the piling up of rubbish and refuse – a source of the spread of infectious diseases;
- air pollution in and near cities caused by traffic, industry and private homes, which is responsible for many deaths;
- calamities such as flooding, tidal waves, and factory explosions, maiming people for the rest of their lives.

"Further away from the cities, other environmental problems can be isolated which are caused by urban living, some of which only become apparent at a later date. These include:
- a change in climate which is the result of the combination of air pollution, the re-routing of the flow of water, erosion, and damage to vegetation;
- dessication, due to the exhaustion of ground water resources (the water is used for both drinking and industrial purposes);
- damage to natural diversity because water, raw materials and energy are transported to the city, or because the flow of wastes from the city contains non-degradable material which, in turn, results in the death of animals and plants."

Everyone knows the problems of the "mega-cities", of the slums without sewerage and clean drinking water, where there is no water drainage and whole districts are flooded during the wet season. People living in these districts are caught in a vicious circle. Their need for a most rudimentary income is so great that they can be easily exploited. The victims have neither the time nor the means to take the most elementary care of their district. Because of the low level of

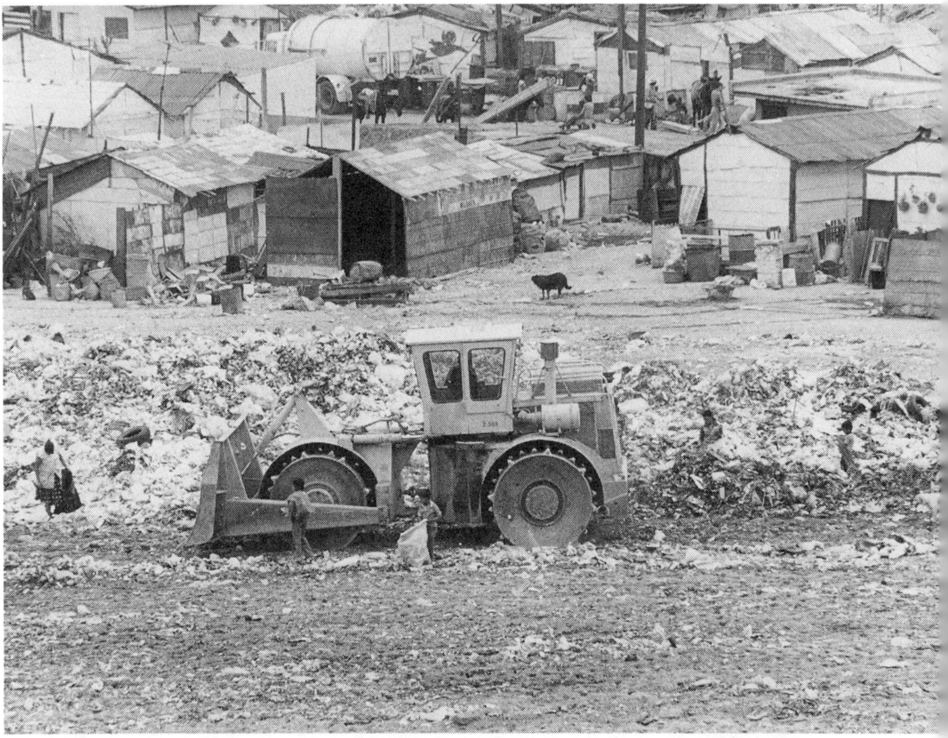

The subject of this photo is not as depressing as it seems. People in many city districts throughout Mexico have organized their own self-help sanitation units, and are cleaning up public areas. photo: Radtke/Present

hygiene, many people become ill and this affects their productivity. The sick people, moreover, often need expensive medicines.
Elementary goods such as firewood and food are scarce and therefore expensive in a rapidly growing city. Because legal entitlement to the ground in slum areas is uncertain, every investment in home improvement becomes a risk. In the degenerating living environment, social norms decline, especially among the young. The growth of destructive and aggressive behaviour often flourishes under these kinds of conditions.

The growth of the cities is explosive and impossible to stop. No longer are farmers, deserting the poor country areas in search of a better life in the big city, responsible for this process. The city itself is growing and the city-dwellers themselves contribute the most.
We can't curb the process of citification. We have to concentrate on looking for ways of making the city a healthy environment.

A healthy city involves more than just public transport and clean technology. Deelstra mentions the following[4]:

- decreasing the distance between the workplace and the home;
- using animals for tractive power;
- extending many forms of public transport;
- altering traffic regulations in favour of public transport;
- minimizing the number of places accessible by car;
- stimulating the use of bicycles by creating cycle paths and encouraging various forms of cycle transport;
- guaranteeing the safety of footpaths;
- creating public areas with difficult access for traffic.

In a healthy city it will be possible to practise farming, forestry and animal husbandry. A healthy city will have its own water purification system, rainwater will be used efficiently, and areas will be made available for composting and making biogas. People will have to find work in their own city, in labour-intensive sectors such as farming and forestry. People will be healthier because the sanitary facilities will be good. Food can be grown locally, which will decrease the people's dependence on the market. Less energy will be required, partly because of the implementation of more efficient heating systems and biogas, partly because of local forestry. The demand for water will decline because there will be a more controlled use of water and because the collection and use of rainwater will be more effective. Efficient home construction and an effective use of parks and gardens can help to make the modern city much healthier. Because a number of large-scale measures will have a positive impact as soon as they are implemented, every modern city would benefit by beginning on the changes now.
Healthy cities are possible. They are a prerequisite for the sustainable development of this world. Environmental organizations and development organizations should combine their efforts and work on these

issues in the coming years. The most important source of inspiration in this work will be local groups and neighbourhood organizations. They will need to be given leeway. National and international policy will have to make sure they are given this room.

> "The relationship between population growth, poverty and the environment is complex. The economies of an increasing number of countries are growing less quickly than their population. These countries are increasingly unable to produce sufficient food for their own people, to finance education and health care and to create employment. This leads to poverty." (Pronk, p. 70)

According to *A World of Difference*, the other side of the problem is that

> "a rise in economic growth well above the population growth rate ... has so serious an impact on the environment and natural resources that it too inevitably leads to poverty." (Pronk, p. 70)

This is a well-known dilemma. When people are poor they are forced by necessity to strain the resources of the environment (for example by grazing more sheep than the soil is able to support). They also have a lot of children as an insurance policy for their future. To conquer poverty, the economy has to grow. But economic growth damages the environment to such an extent that it loses its capacity to support life, and people once more become paupers. The Brundtland Commission, an important group of people headed by the Norwegian politician Gro Harlem Brundtland, published the report *Our Common Future* in 1987. This report was lauded by governments, industry and the banking world alike as providing a set of guidelines for the future.

The Brundtland Commission faced this same dilemma, and Pronk does not agree with the choice it made:

> "If anything at all is to be done about poverty, world industrial production must, according to the Brundtland Report, increase five- to ten-fold from its present level over the next 50 years. If this increase in production is to be at all sustainable, the technologies employed must be at least five to ten times more efficient in their use of natural resources. This begs the question: is this feasible? Should limits not rather be imposed on both population growth and the scale of

production in order to ensure sustainable use of the environment?" (Pronk, p. 71)

Within the concept of "strong sustainability", economic growth is a form of welfare improvement whereby the fight against poverty is combined with the protection of the environment.

> "Despite an initial rapid decline in the death rate, an improvement in welfare in the shape of incomes and access to education and health care generally leads to fewer children per family and so to a reduction in population growth. This demographic transition, which occurred in the rich countries decades ago, is proceeding too slowly in many parts of the world because economic growth is lagging behind population growth. Hence the need for an active population policy incorporating programmes aimed at limiting the size of families and forming part of a broader development policy." (Pronk, p. 87)

But this is not enough:

> "Making family planning services and contraceptives available will be effective only if women are liberated from the traditional cycle of working merely to obtain food, water and fuel, coupled with a rapid succession of pregnancies. Institutional and cultural barriers to the full emancipation and autonomy of women must therefore be removed. It is very important that women and girls should have access to education, and that they have an equal right to organize themselves and participate in decisions on the development of the community in which they live." (Pronk, p. 88)

"We have no money, we are not able to grow food for our families, the water is polluted with agricultural poisons, we are often sick and there isn't even a doctor here." These are the words of Sri Lankan mothers who do not understand why the *loku mahatturu*, the big man, decided to set up a colony in the north of their country, similar to the one in which Sunita and the Banda family live. "We are going crazy here," some say. "We don't know anyone here. We do not live here as people, but as wild animals." They find the problems of loneliness and poverty unbearable. The feelings of shame and the sense that they have lost their self-respect are even worse.

Women work at least 16 hours a day in the productive season. Their position is not improving. With the modernization of agricultural systems used in Sri Lanka, the best parcels of land are now used for the production of high-yield crops. Men are responsible for this production. In the eyes of governments and experts, men are also the providers in the family.

The women are responsible for the production of the day-to-day food requirements. This process most often takes place on the least fertile land. Besides this work, women are obliged to work – paid or unpaid – on the fields in which the market crops are grown. The poorer the family, the more important is the work of the woman in producing the family's own food requirements. At the same time, her role as a cheap labourer is of vital importance to the family.

Women have little say in Sri Lanka. The repressive role-pattern dates back to before the colonial period. But in the evolution of the traditional society of Sri Lanka, women had recovered some space. All sorts of mechanisms had been created to protect women from the worst excesses of the unequal male/female relationship. One of these was the tradition that, on marrying, husbands moved to the land or the village of their brides. In this way the woman was able to rely on her family for support if marriage problems arose. Another mechanism was the division of labour in the fields. Women were responsible for shifting agriculture. They created self-sufficient working groups which formed part of an important social network; they supported each other and shared information.

These mechanisms no longer exist in the colony in which Sunita and the Banda family lives. Everyone has to work for money and the women have been delivered into the hands of the men. The original village has been scattered over many different colonies, so none of the women has her family living close by.

In her book *Mothers for Life*, Joke Schrijver describes the situation of Indra in northern Sri Lanka:

> "Indra is a member of a poor family and is married to a man who has no land and who works as a casual labourer. They have a hut with only one room. Together with her three small children, she is dependent on his wage, but she is never certain whether he will give her money or how much. Indra is always either pregnant or breast-feeding; being responsible for the care of the children, she is unable to obtain work. Her husband forbids her to take the pill and she is unable to use an IUD because she does not dare to go alone to the clinic in the town. If she were to take someone with her, her husband would learn about it. Her third child died shortly after it was

born and shortly afterwards she was again pregnant. Indra's husband blames her for always being pregnant. It is increasingly difficult for them to manage and they quarrel continually. He is very hot-tempered and easily inclined to hit her.
"Shortly after the fourth child was born she ran away with the children, but her husband came after her, hit her until she bled, and took the infant daughter away. Once again she fled to her parents, who lived in another village, but there too the food was insufficient for all the extra mouths. When her father thrashed her mother because she wanted to shelter their daughter, Indra returned home. On the last occasion, she had returned because her husband had threatened her relatives with a knife to force her to go back to him."

3.3 The greenhouse effect

"One of the major challenges for the coming decade and the ensuing period will be to determine how the world can cope with the greenhouse effect and its climatological implications. The atmosphere has undergone substantial changes in the course of the earth's history. The speed at which climatic changes are now occurring is unprecedented, however. If policies remain unchanged, the rate of change will be many times greater than in the past 10,000 years." (Pronk, p. 72)

Everyone knows the phenomenon of the lull before the storm. Heavy clouds build up in the distance. In a few minutes' time they'll burst, but for now it's as quiet as a whisper. The first twigs begin to rustle, the first drops fall. And then, suddenly, heavy rain pelts down and the storm rages. The greenhouse effect and the discussions on the greenhouse effect are in many ways similar to this phenomenon. The weather we will have in ten or fifty years' time is unreal and unpredictable. One thing is certain, though. Rising levels of greenhouse gases in the atmosphere – gases which trap some of the sun's heat which is reflected by the earth and would normally escape back into space – will change the earth's climate quickly and possibly irreparably. We are now in that period immediately preceding the visible change. The extremely cold winters, hot summers, unprecedented storms and rains herald the next stage.

There are three categories of substances responsible for the greenhouse effect. Half of the climatic change is due to the use of fossil fuels such as coal, oil and natural gas. These are responsible for most of the emissions of carbon dioxide, the most important greenhouse gas. Fossil fuel burning also results in emissions of methane, nitrous oxide and the gases which lead to the formation of low-level ozone. These are all major contributors to the greenhouse effect.

A second category of emissions is released by changes in land use and farming practices. Carbon dioxide and, to a lesser extent, methane are released through the burning and clearance of forests and grasslands. Methane is also produced in large amounts by rice paddies and by cattle and other livestock. The sources of nitrous oxide are not well accounted for, but a significant proportion is thought to be released from chemical fertilizers. Land use practices are probably responsible for around 35% of global warming.

Table 2. Relative contribution of gases generated by human society to the greenhouse effect, by source (in %)

Source	CO_2	N_2O	CH_4	*CFCs*	*Other*	*Total*
Fossil fuels	43		3		7	53
Deforestation/land reclamation	18		2		1	21
Wet rice cultivation			5			5
Chemical fertilizers		4				4
Solid wastes			2			2
CFCs				12		12
Cattle farming			3			3
Total	61	4	15	12	8	100

Source: RIVM, 1990

A third category consists of chlorofluorocarbons (CFCs) and related synthetic gases which are mainly used as coolants in refrigerators and air conditioners, as blowing agents in the manufacture of plastic foams and as propellants in spray cans. CFCs are, molecule for molecule, the most powerful greenhouse gases and are also leading to the global depletion of the ozone layer. The actual impact on climate of CFCs is complicated by the fact that the loss of ozone in the stratosphere is now thought to be leading to a cooling of the earth which may equal the extra heat absorbed by the CFCs. The CFCs' indirect cooling effect on climate may therefore be cancelling out their direct warming effect.

The United States, Canada and the whole of Europe are responsible for 65% of the emissions of carbon dioxide and 78% of the emissions of CFCs, Asia (excluding Japann) accounts for 21% of all carbon emissions and 41% of all methane emissions in the world. Latin America is responsible for 5% and Africa for 3% od world carbon dioxide emissions. It is important that these ratios are known and understood, because the countries which discharge the gases must be forced to implement appropriate measures. The more gases a country emits, the more it will have to reduce these emissions. Though warnings about the greenhouse effect have been sounding for many years, the current UN-process to create a climate treaty only started after the first reports of the International Panel on Climate Change were published in the mid-'eighties, while the actual negotiations only started in early 1991.

Pronk states:

> "It is unrealistic to try to achieve a situation in which no climate change takes place at all. That would imply an almost immediate end to the fossil fuel era. The most that can be attained in the medium term is the stabilization of concentrations in the atmosphere at their present level. This would mean completely phasing out CFCs and reducing CO_2 emissions by at least 60%. For industrialized countries this would entail, given proportional distribution, a far greater reduction: ultimately, by about 80% in the Netherlands and 90% in the USA." (Pronk, p. 89)

Pronk's analysis is similar to that of the environmental movement. Two Dutch environmental organizations, Friends of the Earth (Vereniging Milieudefensie) and the Nature and Environment Society (Stichting Natuur en Milieu), say that CO_2 emissions must be reduced by at least 85% in the Netherlands[5]. The Dutch government is not willing to consider a reduction of more than 3% by the year 2000 (relative to levels in 1989)[6].

The wealthy Western world has a heavy responsibility in resolving these large-scale international environmental problems.

> "Whether or not an international consensus is achieved will very much depend on the position adopted by the industrialized countries at the negotiations. Not only are current emissions of greenhouse gases far higher per capita in the industrialized countries than in the developing countries: in cumulative terms the rich countries have also made the greatest demands on environmental capacity. The pattern they have followed in achieving their present welfare is no longer open to developing countries. Unless rich countries begin to be serious about their own policies, the developing countries will not be prepared to subscribe to international agreements on problems which they believe have their roots in the North. Efforts to persuade developing countries to conserve tropical rain forests will easily be seen as hypocrisy if acidification leads to the widespread destruction of forests in the northern hemisphere." (Pronk, p. 88)

The free market and money

It will not be very easy to tackle the greenhouse effect. For one thing, the free market system is a hindrance. Secondly, to transform our economy into a sustainable one will require enormous financial investments. It has been calculated that a 50% reduction in CO_2 emissions will cost approximately US$500 billion a year. Of course, if environmental values were to be included in this calculation the investment would certainly prove to be cost-effective. These calculations should be handled with care. They are based on highly dubious estimates. Some investments, like many of the investments in energy-efficiency, will prove to be cost-saving. It is in the interests of the industry to present a high bill: firstly because it will make the negotiations on the climate treaty more difficult; secondly because it will re-channel large amounts of money back into the industries themselves. Southern nations play an important role in the financial negotiations related to a climate treaty.

Countries in the South will have to receive support from the North to convert their factories. The amount needed is estimated at US$50 to $100 billion. The question is, how do we raise this amount of money?

One possibility is to introduce a world-wide CO_2 tax. Producers would pay tax on each kilogramme of CO_2 emitted, and the returns would be transferred to a special international fund.

It would also be possible to set emission quotas for certain countries or firms linked, for instance, to the population figures. Countries with a small industrial base will not make use of their entire quota while industrialized countries will probably exceed theirs. Developing countries could then rent out the portion of their quota they are not using to the industrialized countries and invest the income they obtain from this in technology transfer and ecologically sound methods of production.

These are just suggestions – it is interesting to note, however, that ideas of this kind are being discussed and elaborated on in an increasing number of international forums. Pronk argues that whatever system is chosen, the quotas must be just:

> "that is, geared in the course of time to reducing the inequality in the demands made per capita of the world's population on environmental capacity. In all probability this will require redistribution on a far larger scale than is currently the case (if only because welfare inequality on a global scale leads to a permanent race to catch up, which is translated into non-sustainable economic growth). If this requirement is not satisfied, the already laborious negotiating processes will undoubtedly fail." (Pronk, p. 89)

On this point Pronk directly clashes with protagonists of the free market principle. Uninhibited and open competition is the basis of the market, and a rejection of its competitive element and the introduction of correcting mechanisms (CO_2 taxes and the redistribution of returns) will be adamantly resisted by the institutions now dominating the economy.

> We know there is a limit to the amount of carbon dioxide we can discharge into the atmosphere without suffering the consequences. Vegetation and the oceans absorb part of the CO_2, but a major portion remains in the atmosphere and contributes to the greenhouse effect. Oceans create most of the "sink" for CO_2. The well-respected scientific "Intergovernmental Panel on Climate Change" has calculated that the oceans absorb between

1200 and 2800 million tons of carbon dioxide a year.
Critical groups from the Third World and groups from the industrialized countries are presently involved in a debate on the allocations of the earth's capacity to remove the greenhouse gases from the atmosphere. The point at issue is the amount of carbon dioxide and methane a person may produce yearly. Groups from the North argue that countries emitting a large amount of carbon dioxide should have a right to more of the global sinks. The United States, responsible for about 15% of global CO_2 production, would then be allowed to claim 15% of the "sink" for CO_2. Kenya, on the other hand, producing only 0.036% of global emissions, would likewise have a right to 0.036% of the carbon sink.
Third World groups, however, argue that everyone has an equal right to this natural mechanism. In this case the United States, with 4.73% of the world's population, would be allowed to make use of 4.73% of the "sink". Kenya, with 0.48% of the world's population, would have ten times as much capacity as in the first calculation.

This sounds quite academic. The discussion becomes more interesting when money is involved. Calculations have been made to determine the amount of carbon dioxide which can safely be emitted on a global level (without causing a dramatic greenhouse effect). According to these figures, we are allowed to produce 300 billion tons of carbon in the form of carbon dioxide between 1985 and 2100. This amounts to approximately 2.6 billion tons a year. Other, more optimistic, scientists arrive at a safe level of 4.8 billion tons of carbon a year.
Sri Lanka presently produces only 3 million tons of carbon in the form of carbon dioxide. If, according to the experts from the Third World, people of Sri Lanka have the right to produce as much CO_2 as the Dutch (within the restrictions of 4.8 billion tons per year for the entire world), Sri Lanka with its 0.33% of the world's population would be allowed to produce an extra 13 million tons of carbon dioxide per year. The Netherlands, however, with only 0.28% of the world population, would have a right to only 13.5 million tons of emissions a year, instead of the 36 million tons it produces at present.
Now we will go a step further and assume that carbon dioxide production permits have not only been introduced but that they are also transferable. If the Netherlands wished to continue producing carbon dioxide above its quota it could buy unused quotas from non-industrialized countries. In the case of the situation just described, to acquire Sri Lanka's unused quota of 13 million tons, the Netherlands would have to pay Sri Lanka an amount of US$15/ton (this rate is often mentioned as being realistic) amounting to

nearly US$200 million. With this money Sri Lanka could invest in alternative energy sources, energy conservation and ecologically sound activities.

An interesting proposition, but it has its problems. Because the industrialized countries are rich, they would be able to buy up the carbon dioxide quotas of countries in the South. This restricts the ability of these countries to industrialize while it allows rich countries to continue polluting the environment. Third World organizations have another objection: they argue that a distinction should be made between carbon dioxide emissions resulting from the realization of basic needs, and emissions which are the product of manufacturing and consuming luxuries. The first kind of emission should not be transferable. Two Indian environmental activists, Anil Agarwal and Sunita Narain, explain it this way: "If the rich countries say we are responsible for half of all greenhouse gases, this means that poor countries will have to cut half of their survival activities (cattle raising, use of biomass and production of rice) while the rich will have to cut half their luxury activities (eating meat in air-conditioned restaurants and driving in gas-guzzling automobiles)." A third objection is the fact that the system is based on creating a large flow of funds. Who will be able to control this flow, and make sure that it will be used for a just cause? Isn't there a chance that the funds will give rise to another wave of destructive investments in the interests of the rich?

For the industrialized world, these calculations are a threat. Research groups in industrialized countries have made alternative calculations in which allocated "sinks" are related to carbon dioxide emission rates. In order to decrease the bill that the industrialized North has to pay, estimates have been made of carbon dioxide production in Third World countries which are much higher than in reality.
It is just a matter of time. Whatever happens, there is a bill and the North will have to pay it.

3.4 Tropical rain forests

"The conservation of tropical rain forests is of global relevance: because of their biological diversity (they account for approximately 50% of all species) and because of the effects they have on regional and global climates. The tropical rain forest problem is comparable to the greenhouse problem. The major difference is that the tropical rain forest is located in a small number of countries: three countries – Brazil, Indonesia and Zaïre – account for 50% of the total area." (Pronk, p. 91)

It is five-thirty in the morning. A bug is crawling over my toes; in the moist twilight outside, the forest is groaning and squeaking. I am lying on the floor in a small shed next to a canteen. My mat is soft and made of the bark of an enormous tree. The Puyamkuti River, which flows through the central part of Kerala in South India, murmurs in the background. My guides and I get up, and go for a walk. We follow a narrow path into the forest. Having arrived at an overhanging bush, one of my guides whispers: "There are often elephants here, wait for me." Half an hour passes. Thousands of butterflies cling to the moist rocks and dragonflies flit by. A blue kingfisher darts across the water. "They have calves, we have to wait." Elephants with calves are very dangerous, I'm not even allowed to see them. Later my guides point out the excrement as proof. We spend the day in the deep forest and when we talk it is in whispers. I suddenly see a chicken, a little larger and more colourful than those I know from home. It must be one of the ancestors of our domesticated fowls.
We reach the dam before evening. For a moment we consider sleeping next to the waterfall, but the thought of tigers and elephants helps to change our mind. We sleep in a little house occupied by three men responsible for maintaining the dam. Sticky rice is their daily menu. We catch some fish and have a pleasant meal. The following morning we go to the reservoir. There are tree trunks floating in it: "Rosewood," says my guide, "one of the most valuable products of the forest." The dam, which was built to generate electricity, has attracted a large number of people: labourers, responsible for maintenance, and landless families who have come in search of a livelihood. We see their houses in the distance, the land surrounding them is denuded but for some large trees too thick for simple axes or saws. The water level in the reservoir is low. Much of the rain forest has already been chopped down and the hills suffer a great deal of erosion during the rainy season. In the rainy season, water and mud fill the reservoir and threaten to breach its walls. The water pressure and the mud at the bottom of the lake may one day cause an earthquake – some earth movements have already been felt. A heavy earthquake might cause the dam to break.

Deforestation has also changed the climate. There is less rainfall now. "If it continues this way," my guide says, "there will be too little water in a few years for the dam to function. We will then be left with a mud pool held in by a weak concrete wall. The rain forests are being chopped down by illegal colonists and by commercial foresters who have never even heard of reforestation. Anyhow, it would not even be possible to reforest this area as soil erosion takes place immediately. It will take centuries for grass to grow here. In the lower regions teak plantations, boring monocultures, have been started. After they have been logged two or three times they are exhausted."

We walk around the lake. We see a strip of rain forest here and there. The elephants are still around. Although the hills are largely denuded there is a lot of grass. "The grass burns well. In the dry period fires are kindled or they start by spontaneous combustion. The air gets thick with smoke." The elephants are bathing in the lake, and squirting mud over each other. It is safe to observe them from a distance. But I would prefer not to be able to see them. I would prefer it if they were hidden by thick foliage, deep in an eternal rain forest.

"The threats to the tropical rain forest are largely exogenous. Poorly controlled, non-sustainable commercial exploitation and large-scale development programmes (mining, dams and the associated infrastructure) may not seem important in quantative terms, but they open up the forest to landless and marginalized people who have been evicted from farmland elsewhere. Many of these direct threats to the forests are in themselves the result of a combination of national policy and international economic relations." (Pronk, p.91)

The Dutch government has a policy on the protection of rain forests. The Netherlands would like all tropical hardwood to be produced in a sustainable manner by 1995. Sustainable forestry will be supported by assisting the organizations of the people living in the forests, by a healthy national economic policy which will reduce a country's dependence on the export of hardwood, and by an effective system to control whether production is genuinely taking place in a sustainable manner. The countries with tropical forests will have to formulate policies to reduce migration to these areas (land reform, population control, employment generation). The Netherlands actively supports this forestry policy by supporting the Tropical Forestry Action Plan (TFAP).

Criticism

Central to the Dutch (and, to a large extent, international) forestry policy is sustainable forestry. In 1986 a large number of countries founded the International Tropical Timber Organization (ITTO), which focused particularly on sustainable logging. ITTO initially received a great deal of support from non-governmental organizations such as World Wide Fund for Nature and Friends of the Earth International.

The IUCN, one of the world's most important organizations for environmental protection, stated: "One of the best ways to make use of the forests – a way which, in theory, can save many species of plants and animals – is natural forest management for timber production." Another organization, the International Institute for Environment and Development, takes it a step further: "Sustainable management of natural forests for wood production is one of the keys to saving the forests and the lumber trade."

This has all proven to be wishful thinking.

Tropical forests are very vulnerable. Sustainability excludes their commercial use. As a specialist working with the African environmental department of the World Bank put it: "From the point of view of the forest eco-system, no form of commercial forestry has ever proven to be sustainable in tropical forests. Commercial forestry can best be compared with mining: it is unable to conserve the environment." This discussion has led to controversies within the environmental movement. While the business community and politicians are proving willing to involve "sustainability" in their commercial priorities, it is becoming increasingly more apparent that it is the commercial activities themselves which have to stop. At this point in the discussion, we are confronted with major interests. Pragmatic environmental activists and the business community have reached a compromise. The question is whether they aren't throwing away the rain forest with the bath water.

4. Narrower margins

> "The turbulent developments of the 1970s and 1980s have clearly demonstrated a growing interdependence between countries. This has been enhanced by the increasing internationalization of private industry. The need to increase the scale of economic operations led to a growth in cooperative undertakings, while the drive to cut costs led to a continuous search for cheaper business locations.
> "The consequence of the developments which I have briefly sketched has been that the scope, except perhaps for the very largest nations, for individual countries to pursue an independent policy has been greatly reduced." (Pronk, p. 103)

Most countries have to operate within margins which are becoming increasingly narrow. Pronk is optimistic about the perspectives of countries which operate within these narrow margins:

> "Developing countries are orienting themselves more towards international markets than they did in the past. Their interests must be safeguarded, albeit in a different way from in the past: it is more important that market access be improved than that the possibilities for developing countries to shut themselves off from the world market are increased." (Pronk, p. 103)

4.1 Underlying factors

The events of the past decade are portrayed

> "as a continuous process of development in the direction of progressive industrialization and internationalization of the world economy. The importance of commodities has declined, that of industry has increased. This applies both to world trade in general and to exports from developing countries.
> Two factors – the oil price increases and the debt crisis – have interrupted this long-term trend for developing countries, but without

> fundamentally deflecting it. On the contrary, the first oil price rise, though initially leading to 'commodity euphoria' among developing countries, in fact helped to accelerate the reduction in dependence on commodities. The debt crisis increased the pressure on developing countries to pursue development through industrialization aimed at the world market." (Pronk, p. 109)

At some point, a choice was made to increase exports. Theory teaches us that the reason for this is that developing countries are need goods in order to develop. These goods have to be imported, for which the countries need foreign currency. Foreign currency is acquired through overseas sales – in other words, through exports. Some of the goods needed for development can be produced domestically. In this way, a country does not have to import as many goods. This strategy is called import substitution.

> "A country that opts for a policy of promoting exports, rather than import substitution, must orient itself resolutely to the outside world and liberalise its domestic markets for goods, services and capital, and its imports and exports themselves. This will produce a higher rate of economic growth than a policy which, with the aid of import substitution, aims primarily at production for the domestic market. The foreign market must, however, offer greater opportunities than domestic demand (for example because the domestic market is small or the level of disposable income among the population is low). Moreover, the policy will be more successful if the proposed exports do not require too many imports (of components, for example)." (Pronk, pp. 109/110)

Trade, it appears, has been declared sacred by the experts who wrote *A World of Difference.* While Pronk had previously recognized the dilemmas relating to economic growth and distribution, he now states:

> "A policy geared to the world market is, in principle, a neutral strategy, so that in certain sectors production for the domestic market will have the upper hand, while in others production for the world market will predominate. The balance depends on the comparative advantages and disadvantages which a country possesses in particular sectors. These may alter over time, bringing corresponding changes in the structure of output and sales. The important thing is to move with the times as

Hong Kong, the world's third largest financial centre

photo: Vivant Univers

regards the industries to be supported, in order to avoid falling into the *once an infant, always an infant* trap." (Pronk, p. 110)

This statement needs further examination. Does it make sense? According to Pronk, trade is neutral and the best way to generate income is to follow the market. *A World of Difference* is full of unresolved arguments. This is one of them. In contrast, Pronk had stated previously:

"The adverse effect on culture consists primarily in the displacement of indigenous views on well-being and poverty by a more materialistic view." (Pronk, p. 78)

He rightly notes that our world-view, with its corresponding value system, is a product of Western civilization:

> "Of relevance in this context are the image of humanity outside nature, of humanity as the master and manipulator of nature, the denial of the intrinsic value of nature, and the idea that well-being depends on constantly increasing material welfare. The radical changes of policy that will be needed in the 1990s must be accompanied by a change in cultural value patterns. There is perhaps a great deal to be learned from what remains of non-Western cultures." (Pronk, p. 82)

Surely this statement conflicts with the earlier assertion that the battle for existence in the market place is neutral? Pronk describes several mechanisms that led to the internationalization of trade. The increase of scale and technological evolution are strong determining factors in this process.

> "The use of modern communication systems (and information technology) to improve access to information has revolutionized the financial markets. Capital markets are continuously in contact with each other around the world. The consequences of this for exchange rates, share prices and interest rates are constantly apparent." (Pronk, p. 111)

Lowering the barriers to monetary trade strengthens the international market immensely. By using a similar form of high technology it is now possible to minimize stockpiles and to anticipate the needs of a manufacturer, which means that raw materials and semi-finished products can be supplied precisely when they are needed. These "just-in-time" deliveries favour large companies that have the capital to invest in transportation and communication systems. The role of the raw materials industry is declining as a result of changes in technology. New materials are constantly being introduced to the market. By improving the efficiency of fuel consumption, lighter cars can now be manufactured. Raw materials are generally being more efficiently used. To be able to participate in this market, companies have to be well-funded, to perform advanced research and maintain flexibility. A production line quickly becomes obsolete under these fast-moving conditions. This is not a climate in which smaller companies, which rely on a longer investment period, can compete. Their factories need to get a return on investments within two or three years.

Shoe-maker in Nairobi, Kenya . How can his trade compete with the massive plastic shoe trade? photo: Bijeen

The first to suffer from internationalization are obviously the environment and the local population. Small-scale enterprises lose their markets; workers become dependent on insecure work (the employer can relocate the factory if pay rates are lower elsewhere); the mobility of the larger organizations makes it difficult to lay down standards for environmental pollution (if the environmental regulations become too stringent the polluters can will simply relocate).

4.2 Are there advantages in exporting?

> "Experience has shown that countries which switched at an earlier date to a development policy geared more to the world market – sometimes as long ago as the 1960s – have been better placed to weather the external crisis. These countries generally enjoy a higher rate of economic growth and export growth." (Pronk, p. 110)

Pronk refers here specifically to many Asian countries and, to a lesser extent, Latin American countries. Africa has missed the boat.

> "Africa still has a long way to go in this respect. More than any other, this continent is in the midst of a process of adjusting to the deterioration in external circumstances. Industrialization has only just begun in Africa." (Pronk, p. 114)

It is true that there has been a gigantic growth in exports from countries like Taiwan, Singapore, South Korea and the Philippines. This has led to a higher level of consumption and the Westernization of their cultures. Pronk appreciates this. But he forgets that in the long run, Westernization could lead to the demise of the planet. As the number of people who emulate this form of development increases, the ecological margins will become smaller. In time the path will become so small that only the strongest will be able to survive. Economically speaking, there will only be enough space for the largest and most advanced businesses, all of which will have head offices in the North. Ecologically, there will be space only for a marginal society, plagued by pollution and impoverishment.

Plastic shoes on the market in Manila, the Philippines photo: Herzog/Present

4.3 The bankrupting of labour-intensive industry

Pronk does not believe small-scale, labour-intensive, and capital-extensive organizations are able to function in the world market. He does, however, value small-scale activity:

> "An improvement in employment opportunities in the cities, and hence in the income position of the urban poor, will in the foreseeable future continue to be based on the expansion of labour-intensive, small-scale business and informal activity." (Pronk, p. 171)

The local business community is in many cases the backbone of the economy. This is where the majority of the employment opportunities – the sources of income of the population – can be found. To stimulate these employment opportunities, Pronk believes that

> "structural reforms are needed in the fiscal, economic and financial sphere, and the choice of technology must be reassessed in the light of employment considerations." (Pronk, p. 171)

A product of present technological developments is a strong international business community which efficiently crushes, in fact exploits, small-scale businesses. This vertical and horizontal integration (inherent in the free market) leads, in the end, to expansion and to the removal of small-scale, self- sufficient businesses. Small-scale suppliers, and other businesses which depend on the international market, are vulnerable. A stock-market crash, new developments in the field of bio-technology or certain economic policy decisions made by a parent company, can have disastrous effects which can be felt in all corners of the world.
"Capital moves to the place where the relationship between risk and yield is the most favourable", according to Pronk.

There are many mechanisms which could support small-scale, labour-intensive businesses. The first step is for a government to drastically change its policy on providing grants to large industries. It should also introduce levies on those imported goods which force local small-scale businesses out of the market.
Several other policy aspects need careful consideration. One of these is "tied aid". This is the system in which companies based in the donor country are subsidized to invest in the Third World. These subsidies are part of the donor's development funding. Local companies face bankruptcy because of this measure, but the donors seem to take no notice.
Loans with favourable conditions, so-called "soft loans", have been set up so that Third World countries can buy from Dutch companies. These products are often not what the local users need or can use. The Ministry of Development Cooperation contradicts its own goal of stimulating the self-sufficiency of local private businesses by granting these loans. The bulk of Dutch financial aid is given to (semi-)

governmental organizations, which often creates a threat to the local economy. Reports of Departmental field inspectors bear witness to this assertion. In Indonesia, for example, local companies were willing to construct bridges, but financial problems forced them to leave the work to Dutch construction companies. The Dutch companies were supported by Dutch development funding.

5. Agriculture and the environment

5.1 Hunger

One of the first World Food Conferences took place in Rome in 1974. The Sahel region of Africa was in the grip of a huge drought. We had only just seen the end of the first oil crisis. Expectations were high. There were ominous clouds on the horizon. World population growth rates, combined with the dearth of fertile land, an undependable climate and advancing deserts, foreshadowed a disaster of unknown magnitude. In his opening speech, Dr. Henry Kissinger promised: "Ten years from now, no man, woman or child will go to bed hungry." The United Nations Food and Agriculture Organization (FAO) adopted this as its slogan and started a series of ambitious programmes meant to boost food production and eradicate hunger. Seventeen years later more people die of hunger than ever before. Environmental destruction is happening on a scale we never dreamed possible and the prospects for food production are even worse.

5.2 Export-oriented agriculture

Why did the FAO fail to realize its goal of eradicating hunger within ten years? Is it due to the ignorance of farmers, who do not cultivate their land properly? Is it caused by gaps in our scientific knowledge, which prevent us from making effective use of new fast-growing plant varieties? Or did we not have enough money to invest sufficiently in the "Green Revolution"?

According to experts from the Third World and the environmental movement, it was not because of any of these reasons. The FAO policy itself was responsible for its own failure, they say. The FAO strategy was directed towards increasing production through modern agricultural methods, irrigation, and large inputs of chemical fertilizers and pesticides. This approach has pushed farmers into the market economy. Instead of cultivating food for their own use, they now produce rice or tapioca for the European Community, Japan or the United States. The returns on their products are never stable and always too low. To be able to compete, a small farmer has to invest large amounts, and families are often deeply in debt.

Many countries import a variety of very cheap products from the EC. As the EC suffers from over-production and very often dumps the surplus in the South (sometimes even in the form of "aid"), prices on the local market slump. The price of many agricultural products has sunk so low that local farmers can no longer earn a living from the land. The farmers have their own survival strategy: they in turn exploit the environment –through over-production, by setting more and more cattle to graze on less and less land, by cultivating the steeper slopes, and by reducing the length of fallow periods. The result is erosion and desertification. And as the boundary of the desert advances, the poor farmer has no choice other than to move along with it.

New plant species and pesticides increase the value of the most fertile lands. In these areas agriculture has been taken over by powerful and rich farmers: large landowners who produce for the export market.

This change in land-use does not happen overnight. Rules and regulations, low market prices, the dumping of food-products by the North, and debts which farmers incur due to investments in fertilizers and seeds, force farmers into export-dependency. The demands of the importing countries often necessitate a new wave of investments, and many farmers eventually lose both land and livelihood. It is partially because of this process that Third World farmers are reluctant to trust Northern traders – even those traders who are involved in the solidarity movement and who want to service the consumer preferences of ecologically-minded Northerners. If the farmers are asked to produce without using pesticides, they tend to regard the request as yet another trick. They know that many small farmers lost their incomes and their lands in the past because of the tricks of the Northerners. Fertile regions no longer produce food for the local market, instead they generate money which enables the élite to buy luxury consumer goods.

Most forms of intensive agriculture pollute. Thousands of people die each year because pesticides poison ground water and surface water. Irrigation is used to achieve high rates of production. This water, like all water, contains salt. When the water evaporates, the salt remains, and after several years a salt layer covers the once fertile soil. Because the irrigation water is not adequately drained off, the level of the ground water rises, and in the end the land drowns. Each year, the amount of land which is lost in this way equals the surface area of newly irrigated ground!

5.3 The Netherlands as an example

For many Third World and environmental organizations, the way the Dutch practise agriculture is a perfect example of how not to do things. Agriculture in this tiny country (which is, in absolute terms, the world's second largest agricultural producer) suffers from four major disorders:

- Acidification: The enormous number of cattle in the Netherlands produce mountains of manure that cannot be got rid of. Ammonia now permeates the rural areas. The Dutch are slowly becoming aware of the fact that this rural smell is part of the mechanism which is killing their forests. Ammonia not only ruins the soil; in some areas it is the number one killer of forests.
- Over-application of fertilizers: The Dutch inundate the soil with chemical fertilizers. Many of the phosphates, nitrates and other chemicals end up in the ground water and surface water. There is not a single part of the Netherlands where this has not yet happened.
- Dehydration: Because the industrial farmers want the grass to grow early in the spring, because they want to drive heavy tractors through the fields, and because they want to remove salt water, water is continually being pumped out of Dutch agricultural land. The result is soil dehydration. Dehydrated soil burns up, very slowly. It loses its humus. The first eroded and unproductive Dutch polders have now been put up for sale.
- The use of poisons: Dutch industrial farmers apply more poison per hectare than any other farmers in the world. The use of

pesticides in potato, flower and flower-bulb cultivation is world renowned, but large quantities are used on other products as well. Although the most dangerous kinds, such as DDT, have been outlawed, more and more poison is accumulating in the ground water. It will not be long before the Dutch will have to use expensive carbon filters to purify their drinking water.

5.4 Modern agriculture resembles mining

The finite nature of present-day agriculture closely resembles mining: there will come a time when the soil will be so exhausted, polluted or dried out that it will be rendered useless. To continue applying chemical fertilizers and pesticides can do no more than postpone that moment.

It consumes a great deal of energy to produce fertilizers and pesticides – and modern agriculture already devours so much energy itself! Agriculture used once to be based on solar energy. The composition of the soil was improved by adding manure which was generally produced by animals kept on the farm. The combination of manure and plant remains maintained soil fertility.

The composition of soil is now regulated by adding plant nutrients produced in chemical factories. This kind of agriculture is not sustainable. The soil actually degrades, and more and more external inputs are needed if the level of production is to be maintained. Modern agriculture generates high levels of production and welfare in the short term; in the long term it offers no prospects whatsoever.

5.5 Alternative agriculture

Third World groups and environmental organizations advocate another kind of agriculture, one that can guarantee an adequate supply of food in the long term as well as in the short term. Their starting point is that agriculture must become less dependent on chemical fertilizers and pesticides. This "alternative" agricultural system depends only minimally on external inputs, which is why it is known as "Low External Input Agriculture" (as opposed to "High External

Input Agriculture"). Of the food produced in the world today, 80% is produced using this "alternative" form of agriculture. Export-oriented industrial agriculture takes care of the rest.

According to the FAO, alternative agriculture cannot produce enough food to feed the world's population. From its point of view an ecologically sustainable system is ideal, but inadequate.
The productivity per hectare of alternative agriculturalists in the Netherlands is on average 60% that of industrial farmers. If the alternative farmers are successful in allowing the soil to regenerate to its natural state of vigour, the production capacity will increase. This is not a fast process: after having been subjected to industrial agriculture, it will take the soil about five years to recover. Production will further increase in this period by another 20%.
Little research has been done on alternative agriculture and methods of increasing the productive capacity without exhausting the soil. More research would probably bring about new prospects for alternative agriculture.
All in all, it is unlikely that people will go hungry if Dutch farmers switch over to alternative farming methods.

5.6 Agriculture directed towards recovery

Opportunities for alternative agriculture depend very much on local circumstances. Often, because of the methods previously used in agriculture in that area, the circumstances are far from good. Farmers will need a period of transition before they can implement sustainable agriculture. Farmers may need extra support. Their degenerated lands may need some chemical inputs, temporarily, and the farmers themselves will probably need to learn a number of new techniques. In the meantime, the prices for agricultural products will have to remain high enough for farming to be cost-effective.
Although it is likely that alternative agriculture will be able to produce enough food to meet domestic demand, less food will be available for export. Agricultural production will primarily meet the demands of the domestic market. This means – and this is a big problem – that countries will have to be free to produce for their own domestic

market. This can be a problem in developing countries, where, partly because of foreign debts, there is a tremendous pressure to export food. At the same time, policy measures will have to be taken to protect these countries from cheap agricultural imports.

5.7 Stimulating measures

Governments should do more to encourage sustainable agriculture at the consumer and the producer levels. The European Community and the FAO should do likewise.
Incentives are needed in the area of pricing policy. Environmental groups and Third World organizations argue that the price of a product should be based on real costs, which include the cost (damage) to the environment and the cost of the hardship it has caused to people. Agricultural products which exhaust the soil have to be more expensive; the price of the energy used to transport products over long distances or to grow them in greenhouses has to be increased. The same goes for the price of chemical fertilizers. Environmental pollution should be taxed, and pesticides – especially the most dangerous ones – prohibited. The result of all these measures will be a substantial increase in the price of many agricultural products.
Finally, governments should give research, marketing and management in the field of alternative agriculture much more financial support. Marginal areas should receive priority in this regard, as this is where the world's poor population lives.

> The Dutch consultancy firm Berenschot published a report in 1990 containing a cost-benefit analysis of a Dutch change-over to biodynamic agriculture. Berenschot concluded that the prices of agricultural products are presently too low to make biodynamic agriculture economically viable on a large scale. It would be stimulated tremendously if these prices were to change through the implementation of taxes on pollution and a reduction of taxes on labour. A price policy would help considerably in attaining important environmental goals, while creating employment opportunities. It would result, in other words, in more nature and more employment – but we will have to pay for it.

Diagram 1. The nitrogen balance of biological agriculture (assuming it was to be practised by all Dutch farmers), in tonnes/year

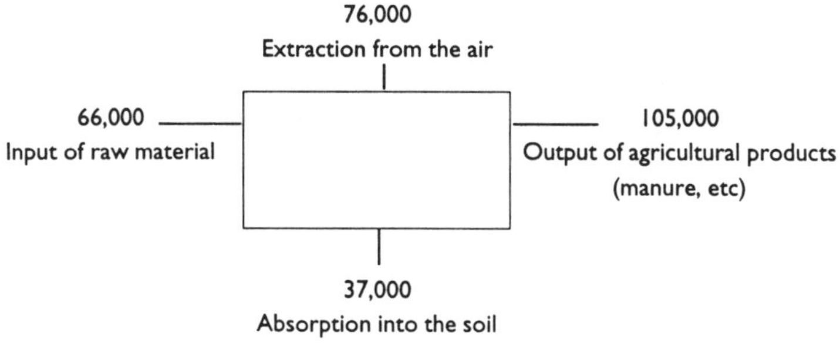

Source: Berenschot 1990[7]

Biological farming consumes 66,000 tonnes of nitrogen from external sources; "smart" bacteria and fungi in the soil extract a large amount of nitrogen from the air; 37,000 tonnes is washed into the soil; and 105,000 tonnes is converted into food. It appears as if this kind of agriculture binds more nitrogen than it consumes.

Chemical agriculture functions oppositely:

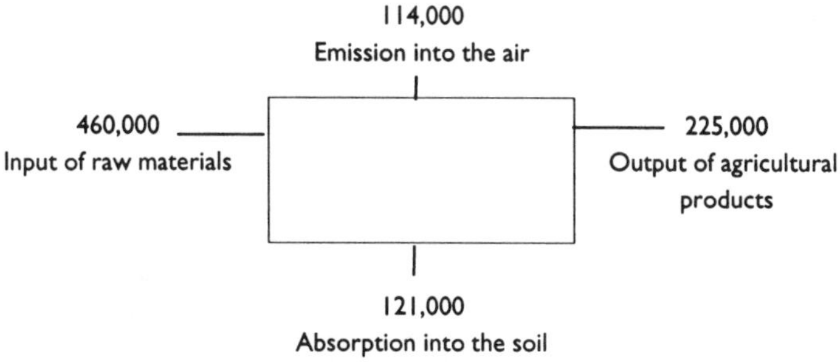

Source: Berenschot 1990

The report also presents an interesting comparison of the "nitrogen balance" of biological and modern chemical agriculture. The diagrams below indicate how much nitrogen is used in the two kinds of farming (for instance by applying chemical fertilizers) and how much the farm generates at the end of the production cycle (in the form of acid rain or food).

In modern agriculture, 460,000 tonnes of nitrogen is used to produce 225,000 tonnes of food. Instead of binding nitrogen, the modern system loses it in almost equal proportions to the soil and the air. As a result, trees die and our drinking water is poisoned. The system is very inefficient: more is put into it than is eventually produced.
In other words, investment is higher than returns. This may be acceptable for a few years, but it is damaging to the environment, and over-production forces agricultural prices downwards, putting many farmers out of business.

The FAO pumps stimulants into highly productive agricultural regions. It suggests to these regions, not very convincingly, that they should be concerned with the care of the environment. Apparently the FAO is (still) afraid of thwarting the interests of large producers of and traders in pesticides and chemical fertilizers, the seed industry and agribusiness. International environmental organizations and Third World groups will have to continue criticizing the FAO and inundating it with policy alternatives.

Jan Pronk supports Low External Input Agriculture; but at the same time, despite knowing that the success margins are limited, he focuses his attention on the export market and the conditions it imposes on agriculture. The case of Sub-Saharan Africa:

> "The prices of agricultural products on the world market are so low that they do not leave national governments and farmers enough room to invest in sustainable agriculture. This situation is unlikely to change in the near future, given the advances in biotechnology in the North. The international community will have to be urged, therefore, to remove barriers to exports of new agricultural products from Sub-Saharan Africa and, if necessary, to agree to Sub-Saharan Africa being accorded a privileged status in international economic relations." (Pronk, p. 224)

The World Bank policy – which Pronk subscribes to – is based on support to the agricultural sector through financial inputs, research and infrastructural improvements. In this, there is a tension between export-oriented agriculture and agricultural production for the local or regional market.

In an Africa Seminar, organized by the Dutch Ministry of Development Cooperation in 1990, it was argued: "If food production is reduced because of the expansion of export crops, food prices will generally increase. This does not necessarily mean that there will be less food available, particularly if the foreign exchange earned through exports is used to buy foodstuffs. However, there is no empirical evidence of this happening. In a group of thirty Sub-Saharan countries, food imports did not increase when national food production declined. As a result, there was a reduction in the total availability of food."

The large agricultural firms are able to maintain their head start and keep prices low thanks to, among other things, the rapid evolution of technology. This is why there is no prospect of competitive trade in the present products being produced in the Third World. Pronk suggests that the Sub-Saharan region should be stimulated to produce new products, such as vegetables and fruit which grow well in warm climates. Clearly he does not recognize the conflict between this export-oriented strategy and food self-sufficiency

> "Low External Input Sustainable Agriculture (LEISA) is very promising. It features: few external inputs, by keeping cycles closed; lower water consumption; protection of vulnerable soils through permanent biomass cover; wide diversity of crops which ensures, on the one hand, ecological stability and thus the spreading of risks and, on the other hand, an abundance of useful products that meet basic needs (fodder, building materials and fuels as well as a variety of foodstuffs); compatibility with traditional farming systems and the knowledge of them that still exists." (Pronk, p. 95)

In other words, it is essential that connections are made with traditional systems (which is not the same as returning to traditional systems). The export-oriented production of beans, aubergines and other vegetables and fruit for the European market does not make this connection. Yet Pronk chooses the latter.

High and Low External Input Agriculture often compete against each other on a basic level. This is very clear with regard to land ownership. Extensive plantations and cattle ranches producing for foreign markets are generally owned by large firms and big land owners. Most of the world's farmers, however, are tenants or do not even have any land of their own. They sometimes have to engage in a life and death struggle just to obtain even a few hectares of land so that they can meet their own consumption needs.
There are many more advantages that the alternative or traditional sector offers. It offers a greater scope for employment. It supplies other parts of the community with a variety of good-quality products such as energy and building materials. Finally, a factor which should not be forgotten is the fact that the social and cultural appreciation of this form of agriculture is so much greater within the local community.

The Banda family in Sri Lanka used to live in a small farming community. Recently they were forced to move and met Sunita and her mother. That happened at about the same time that they began their resistance to the development experts.
Their community used to have different kinds of agricultural lands. The *chena* was a wooded area suited to shifting agriculture, where herbs, fruit and vegetables were cultivated. Besides *chena* agriculture the farmers also maintained rice fields. Erosion or soil exhaustion was almost non-existent: water, which gradually flowed from rice fields at higher levels to those below, finally flowed into a pool where the buffaloes cooled off. This is a very advanced farming system, the result of a subtle balance between people and nature.

In the past the buffaloes did all the heavy work. They are dark, heavy animals which do not like excessive sunshine and which need to bathe in water so they can keep cool. This is why every village had a pool which always had enough water for the buffaloes to bathe in, even in the dry period. These days, the work of the buffaloes has mostly been taken over by tractors. This is only one of the major changes that have taken place in Sri Lanka's rural areas; the effect it has had illustrates the complexity of the traditional farm system and the dangers of rapid change:

1 The tractor also replaces labour power and reduces employment possibilities. The main victims are landless farmers who were often hired as labourers.

2 The tractor uses fuel and is more expensive than the buffalo (tractors are preferred only because they are a status symbol, even though it costs four times as much to plough with a tractor!).

3 The tractor destroys the soil structure, reducing the soil's water-retaining capacity.

The introduction of tractors has brought these direct changes. The indirect ecological consequences are more far-reaching, and illustrate just how complex the system is.

In the dry season the rice fields are dry and most water tanks are almost dry. Many water species retreat to the buffalo pool which functions as a place of refuge, allowing species to survive the dry season. The woods around the pool and the tank also play an important role in the local ecology. With the introduction of the tractor, people no longer maintain the pool and its surrounding area, and the pool disappears. The pool is no longer there to fulfil its other functions.

4 Fish normally hibernate in the pool and colonize the fertile rice fields as soon as the rains arrive. Fish are an important source of protein for the villagers. They are also a source of income.

5 Many fish – especially those inhabiting the rice fields – feed on insects. They move in just in time to eat the first insect larvae which otherwise spread diseases (such as malaria) and reduce the crop. The disappearance of the pools means an increase in the insect population. The incidence of malaria in regions in which tractors have been introduced and pools have disappeared testifies to this. Pest control using pesticides is expensive and has its own disadvantages: the poisons injure the health of the ecosystem and of the human population.

6 Trees and bushes often surround the pools. They provide a home for all sorts of animals which participate in an important way in the agricultural system. Take the rat snake, for example: it eats about twenty rats a month. This snake is one of the most important aids in managing rat and mice plagues, but in the absence of moist woodlands it cannot reproduce. Other animals also contribute to the maintenance of the system and help prevent plagues.

One last example demonstrates the effect of the disappearance of the water pools to the community.

7 In many parts of Sri Lanka coconut fronds are used to make roofs. Although green fronds are the easiest to process, to pick these fronds would reduce the productivity of the palms. The old fronds, however, are stiff and heavy. To make them more pliable, the fronds are laid in the buffalo pool for a few weeks to absorb moisture. Because the fronds cannot be soaked if there is no pool, coconut fronds can no longer be used as roofing material. The cheapest alternative is roof tiles, but they still cost eighty times as much as coconut fronds. The baking of tiles consumes vast amounts of wood. The Sri Lankan wood supply is being rapidly depleted: in 1956, 44% of Sri Lanka's surface area was covered by forests, in 1976 this was reduced to 22%, and by 1990 only 6% of the area was forest.

Tiled rooves in Gujarat, India photo: Eijgenhuijsen/HH

6. Debts, the market and the environment

6.1 The debt crisis[8]

In the period following the Second World War, the Third World's most important moneylenders were Northern governments and inter-governmental institutions such as the World Bank. This situation prevailed until the 'seventies, when the economic power of the United States became threatened by Japan and Western Europe. In the early 'seventies, the system of free exchange rates was replaced by a system in which the rates of exchange, based on the dollar, were determined internationally. In this global financial system, profits could be made by buying and selling currency in periods of fluctuation.

Private banks started to provide loans to Third World countries on a large scale. The international banking community underwent a rapid expansion. While banks in the United States were kept under strict control, European banks were given more latitude, and the market for European currencies grew. In 1973 the European money market was valued at US$300 billion; by 1983 it had increased to more than US$2,000 billion[9].

Through its activities in 1973-1974 and 1979-80, OPEC (the Organization of Petroleum Exporting Countries) was successful in its bid to vastly increase the price of oil. Arab countries became very rich. They were worried, however, that the major political differences between the United States and OPEC would jeopardize the safety of their capital, which they had invested in United States banks. The Arab countries transferred a large portion of their oil returns to the European financial market.

The leading banking corporations found a way of turning this situation to a greater profit. They bought money on the growing European market and lent it to Third World countries. In this way, large amounts of so-called petro-dollars were loaned to the Third World in the 1970s, often at low rates of interest.

The loans were used in various ways. Military régimes, such as Pinochet's régime in Chile, bought the weapons they needed to suppress their own populations. A large proportion of the loans disappeared into private accounts and very quickly reappeared in the North, as investments or as deposits in savings accounts. The flight of capital from the Third World was enormous: between 1977 and 1987 an estimated twenty billion dollars was transferred from Brazil alone[10]. Cheryl Payer, in *Lent and Lost* (Zed Books, 1991), presents a table which spells out the implications of capital flight for the debt burden, by juxtaposing the actual debt burden for certain countries (at the end of 1985) with the amount of debt they would have had if capital flight had not consumed the borrowed money.

Impact of Capital Flight on Debt US$ billion, unless otherwise noted

	Gross external debt, end 1985		Gross debt as percentage of exports of goods and services	
	Actual	*Without capital flight*	*Actual*	*Without capital flight*
Argentina	50	1	493	16
Brazil	106	92	358	322
Mexico	97	12	327	61
Venezuela	31	-12	190	-55
Malaysia	20	4	103	18
Nigeria	19	7	161	62
Philippines	27	15	327	195
South Africa	24	1	131	15

Source: Cheryl Payer, Zed Books 1991, using Morgan Guaranty Trust Company, *World Financial Markets*, March 1986, table 12, p. 15.

Cheryl Payer quotes the Morgan Guaranty Trust Company which stated that "on the admittedly oversimplified assumption that capital flight serves only to inflate debt build-up and related interest charges, several of the [countries in the table] might have been practically debt-free today – were it not for capital flight". Payer points out that none of the collected statistics on capital flight include interest earned on capital transferred in earlier years, whereas interest is automatically added to debt totals if it is not paid promptly. The actual contribution of capital flight to debt is therefore understated in these tables.

The major part of the loans, however, was invested in industrial and infrastructural projects and was intended to boost economic growth. Due to a number of external circumstances, Third World countries were no longer able to repay their debts.

The second oil price hike, in 1979, placed many Third World countries which were wholly dependent on oil imports in great difficulties. The next shock came when the government of the United States decided to double interest rates to twenty percent. This measure was meant to meet the huge deficit in the US budget by attracting more capital to the United States. The deficit was in part due to an immense investment in the defence industry. The Reagan administration, with its *laissez-faire* politics, reduced taxes in the United States and increased the deficit.

The debt crisis was also caused by the fact that the Third World was earning less on its main source of foreign exchange, the export of raw materials. In the 1980s, the price of raw materials declined on the world market, due to protectionism in the North and to the competitive new synthetic alternatives. Cotton, for example, lost ground to nylon, cane sugar to synthetic sweeteners, and copper to glass fibres. While the income of African countries still increased by 22% per year between 1970 and 1979, there was an annual decline of 9% in the following period (1980-84). Since then there have again been several decreases (0.7% in 1985, 26% in 1986, 6% in 1987)[11].
The Third World's debt problems continued to grow, until Mexico announced in 1982 that it was refusing to repay its loans. A prominent banker compared the Mexican decision to "an atom bomb on the financial market" (*Time Magazine*, 1983). The banks were afraid that the Mexican decision would incite other countries to follow suit,

which would ultimately result in the collapse of the world's financial system and the bankruptcy of many large banks. Governments and banks started working together to prevent the realization of this doomsday scenario. They succeeded in finding funds to rearrange the Mexican debt and in preventing the Mexican economy going bankrupt. The International Monetary Fund (IMF) was given the task of managing both the economic development and the debts of Third World countries, as well as trade imbalances on the macro-economic level.

Third World countries which asked for IMF support in restructuring their debts were compelled to sign a "letter of intent", declaring that they were willing to follow the IMF's economic recommendations contained in a "structural adjustment policy" (SAP). This policy was formulated by a team of economists, which the IMF specifically contracted for each country.

Machines, tractors, and sometimes even whole factories no longer operate due to a dearth of spare-parts. photo: Bijeen

Structural adjustment policies had a number of common elements: the reduction of government expenditure, the stimulation of export production, the reduction of imports, and the privatization of state enterprises. SAPs were meant to ensure that countries earned enough through foreign exchange to repay their loans.
But SAPs also have negative effects on the community in the debtor country. Unemployment rates are exacerbated, the quality of health care decreases, and government subsidies on such basic needs as foodstuffs are reduced. SAPs increase the poverty and misery of the poor.

The IMF has various instruments at its disposal to ensure that debtor countries comply with SAPs. If a country refuses to repay its loans, it is confronted by a number of political and economic measures. These measures include confiscating capital which is invested abroad, confiscating exported goods, the withdrawal of landing rights for national airline companies, the setting up of an embargo on the import and export of capital goods and spare parts, and sometimes even the restriction of food imports.
Countries which do not keep to their debt repayment agreements are also confronted by sanctions. This happened to Zambia, when it was no longer in a position to keep to its repayment schedule, following extensive riots which had been sparked off by sharp rises in food prices. In response, the IMF terminated its financial support to Zambia and the British government annulled a support package for Zambia to the value of £30 million[12].

Third World countries which are largely dependent on the export of raw materials to pay off their debts have few opportunities to rid themselves of their debts. Important decisions, which directly affect the economies of Third World countries, are made in office buildings in the North (primarily by the staff of intergovernmental institutions such as the IMF). We have now entered a new colonial era.

6.2 The relationship between debts and the environment

There is usually a relationship between debts and the degradation of the environment. The IMF's Structural Adjustment Policies (SAPS) compel countries to maintain an export economy, forcing them to over-exploit their natural resources. Several researchers have given this relationship specific attention. The *Freedom from Debt Coalition* in the Philippines asserts that the Philippine foreign debt is repaid by: "destroying our forests to export wood, ruining our coral reefs to export fish, and exhausting our soils by applying heavy pesticides and chemical fertilizers to facilitate export-oriented agriculture"[13]. Ben Jackson, a member of the (British) *World Development Movement*, quotes the example of the Midland Bank, which agreed with the Peruvian government to exchange debts for iron ore and fish[14].

The Brundtland Report, published by the World Commission on Environment and Development in 1987, noted three ways in which the general crisis and recession in the debtor countries has resulted in environmental destruction. First, SAPS have stimulated large-scale monocultures in the Third World, which in turn have forced small farmers to move from their land to marginal areas. Survival agriculture in areas bordering on deserts and tropical rain forests exhausts these areas. This mostly happens when farmers have been forced to invest in fertilizers and pesticides or in new technology which consumes huge quantities of energy, like tractors and diesel pumps. Because the technology is inappropriate or not well introduced, and because of the low price of agricultural products, many farmers could only sustain their farming activities by building huge debts. In the end, their only option was to sell the most fertile land to larger farmers, and move to other areas. Secondly, funds are withdrawn from institutions whose job it is to look after the protection of the environment. Thirdly, in a period of "economic stress" policy makers are inclined to "neglect environmental policy and the protection of nature in industrial and rural development projects".

However, the role of debts in the destruction of the environment in the South is often over-simplified and exaggerated.
The Brundtland Report observed that SAPS have caused governments

to cut back on their funding of the environment. How these cutbacks affect the environment is, however, difficult to trace. Many environmental organizations in the Third World are not very effective, or give their support to development projects which destroy the environment or lead to the seizure of lands belonging to indigenous populations.

The third of the Brundtland conclusions is not easy to quantify. Periodic economic crises are inherent in the global economy, and an insecure future brings pressure to develop at all costs. That is the reason why policy makers are generally inclined to neglect environmental aspects in development projects.

A critical investigation of the relationship between debts and deforestation was undertaken by Peter Madden, of the British development organization Christian Aid[15]. He compared the deforestation figures in thirty countries in the 1980s with their debt levels and repayment rates. He concluded that:

- There is no relationship between the size of the debt and the rate of deforestation.
- There is no meaningful relationship between debt and export, or between debt and gross national product and the deterioration of the forests.
- There is no statistical relationship between the growth rate of debts and deforestation.

Madden's conclusions seem to indicate that the destruction of the environment in the Third World is not only caused by international economic factors, but also by domestic political and economic measures.

The beginning of the devastation of the Amazon forest can be traced back to the end of the 1930s, and the plans of President Getulio Vargas to develop the northern regions of Brazil. The military government which took over power in the country in 1964 designed further plans to colonize the tropical forests. Colonization served a political purpose: to cultivate unity among Brazilians by providing them with a common goal, and to protect the borders by populating the "empty" regions. In 1966 "Operation Amazonia" was announced

The disappearing Amazon forest. Hundreds of tonnes of the powerful defoliant pentachlorophinol were used by the CAPEMI company to clear the way for a high-tension powerline between the Brazilian cities of Tucurui and Belem. Other, unidentified poisons were reportedly also used. photo: RBP

by the president, General Castello Branco. It was immediately embraced by the country's élites, happy to be relieved of the pressure to carry out land reforms.
Towards the end of the 'sixties, the inflation rate in Brazil was very high. Land became an attractive (inflation-proof) investment. Combined with the financial incentives offered by the government to colonize the forest, this led to a wave of speculation in which 60 million hectares of public land was sold to private persons[16].

In 1964 a law was passed which gave farmers the right to use the land. This law encouraged farmer colonists to chop down forests. Speculators often hired private armies to intimidate small farmers to vacate their strips of property. Large speculators deforested immense areas and converted them to cattle pastures, despite the fact that meat production on these ranches was (and still is) not remunerative. The Brazilian Institute for Management and Economic Analysis reported that only about three percent of the large cattle ranches were profitable in the 1980s.

Trees were felled and burned. Ten percent of the Brazilian rain forest (amounting to 350,000 sq. kilometres) has been destroyed in this way since 1970.

Domestic political and economic developments influenced the large-scale destruction of the Amazon forest and the marginalization of the poor population well before structural adjustment policies were applied in 1982. Even now there is not a strong relationship between adjustment policies and deforestation in Brazil. Brazil's exports amounts to only ten percent of the national product. They consist mainly of agricultural products (such as processed soya beans, orange juice, sugar, coffee and cocoa), which are largely cultivated outside the Amazon region.

The debt situation only indirectly reinforces environmental destruction in the South. The exchange rates and the easy terms for loans (because there was so much money) in the 'sixties and 'seventies, caused environmental destruction in the Third World. Large projects which now destroy the environment were conceived in the 'seventies and became operative in the 'eighties. They were not so much forced on the Third World by the North, as actively chosen by local élites.

The money which has caused the Third World to become indebted has been used for a specific style of development – a style which still dominates the world economy. Susan George, in *A Fate Worse than Debt*, calls it "the model of bad development":

> "For many years, in many quarters (and still today), it was assumed that 'emerging nations' had a single goal and must follow a single path to reach it. Never mind the appalling poverty of imagination displayed: people in authority both in the North and in the newly decolonized countries wanted the South to become 'like' the rich, industrialized (often ex-colonial) powers."[17]

6.3 The Brady Plan

According to Pronk, essential elements in the solution of the development crisis are the strengthening of international commerce and trade and the strengthening of the international investment climate. He suggests the following approach to the problem of the flight of capital:

> "Capital will, after all, find its way to those places where the balance between risk and return is the most favourable. The proper way to encourage the return of capital to debtor countries is therefore to create circumstances so that investment once more becomes attractive." (Pronk, p. 129)

A World of Difference does not advocate a thorough remodelling of the structural adjustment policy. Pronk claims that:

> "a lasting balance must be struck between the need for financial and economic reform in the short term and that of preserving and strengthening the longer-term foundation of development: the capacity for growth." (Pronk, p. 129)

Pronk is attracted to the Brady Plan. Briefly, this plan advocates that debtor countries should receive an extra financial injection to ensure that their economies "outgrow" their debts. The Brady Plan sees economic growth as a weapon in the fight against the debt crisis.

Pronk advocates linking debt relief to environmental conservation:

> "The growing realization that drastic environmental protection measures are unavoidable might increase the willingness to accept levies on activities that damage the environment in industrialized countries. Part of the revenue could be used for large-scale debt relief in conjunction with environmental protection in developing countries (*e.g.* protection of the tropical rain forest)." (Pronk, p. 127)

6.4 Debt swaps

Debts can also be "bought up". Banks, realizing that they will probably have to forgo a large part of the money they loaned in the past, are willing to sell cheap debt certificates. Large environmental organizations can buy up certificates and ask the Third World country to repay the debt, not in dollars or any other hard currency, but in local currency or in the form of environmental protection of a certain area – payment in kind. The debtor country is usually capable of participating in this kind of swap: it has enough land or sufficient local currency. Debts can in this way be converted into environmental protection measures.

The idea of debt swaps has come under severe criticism, even though the amount of money involved is small. Some claim that debt swaps are nothing more than a clever use of the debt crisis to obtain more funds for environmental purposes. Others say that the debt crisis is being misused to influence a country's domestic policy. They feel that Western countries and environmental organizations have no right to use this instrument; it attacks the sovereignty of states.

A third criticism is that debt swaps do not tackle the source of environmental degradation and underdevelopment. Moreover, the small amount of money which is released through debt swaps is paltry when compared with the huge sums spent on environmental destruction. Debt swaps divert attention away from the real sources of the problem.

A fourth criticism is that many countries also have a large internal (national) debt. Debt swaps involve freeing local currency for new investment (often for Northern industries). Because of the lack of available currency, the internal debt increases. The government is forced to look to national banks and investors for loans. Alternatively, it can print extra money, and encourage inflation.

6.5 The debts of the Western countries

The North consumes the largest part of the world's raw materials, and Third World countries act primarily as suppliers. The suppliers have conducted an all-out sale: forests have disappeared, fishing waters have been depleted, and enormous amounts of foodstuffs have been exported to the North. Investments, which later became the cause of debts, were used to set up a policy of large-scale extraction. The industrialized world first stole the raw materials, and now it wants its money back!

Many Third World organizations and environmental groups feel that the debt crisis is not being viewed from the right angle. If a country has been ransacked, has it really incurred debts? The North has made a good profit on the South. The IMF's stranglehold is now being used to wring out the remainder.

Various groups in the Third World demand a new calculation of the debt, which includes the costs of the ecological destruction which has taken place. It will not be easy to calculate the total "ecological debt": the devastation of agricultural lands, poisoned air, the threat of extinction to plant and animal species, and emptied fishing grounds. The value of this destruction is greater than the value of the debt. If we are honest, we will admit that the North has a far greater debt to pay the South than vice versa. The countries which have plundered for centuries should now start repaying the damage instead of continuing their thievery.

6.6 A different economy

The debt issue is very complicated. As we have seen above, a direct relationship between environmental damage and debt is difficult to establish, but there are indirect connections. Everyone incurs debts sooner or later. An investment requires money and taking out a loan is not necessarily wrong. Things go awry, however, if a debt keeps mounting and finally results in a bargain sale. This happens when the loan is badly invested, when it is invested, for example, in projects which operate within the context of the unpredictable international market. If ecological consequences are not considered in the planning phase of projects, a loan is badly invested. These projects appear to have favourable prospects when in fact they undermine the environment and with it the country's future. There should be an integration of environmental and social goals at the economic decision-making level. That happens in the North. If a country in the North has to make a decision to invest in a coal-fired power plant, a large number of factors are investigated and weighed – an environmental report is drawn up, employment effects are calculated, the installation is examined and safety aspects are considered.

Increasingly, we try to translate into economic units environmental effects, for example by taxing products which pollute. It is not possible to translate the effects fully: we cannot imagine all the environmental problems we shall encounter in the future. When CFC factories were built, we did not know about the hole in the ozone layer so we couldn't include those costs in our calculations. Similarly, there are likely to be many industrial products today which have ecological effects we will only discover in ten or a hundred years time. We have to make careful calculations.

Money is not the same as environment. Money can only substitute for the environment to a limited extent. We could, for example, accept a certain amount of erosion in a particular agricultural project, and compensate the farmers for their losses. But the farmers lose more than income alone: they lose their land, their culture, their identity. Much of this cannot be expressed in monetary terms.

Finally there are products which should not be produced at any price. CFC, cadmium, DDT and other extremely harmful substances should be prohibited altogether.

Our economy has to take environmental effects into account. This can be done in various ways: by imposing duties, by targeting investments in ecologically sound sectors, or by reducing taxes on labour and increasing taxes for pollution. To carry out these and similar measures, laws must be made and enforced at a central level. This will create tension between the state, which thinks up the rules and monitors compliance, the individual and the business community.

The challenge is to organize this jurisdiction in such a way that producers and consumers are part of both the decision-making and the monitoring systems. In this way, the people and not the centre will be in control.

Ricardo, who is from the old hacienda Lampa close to Santiago, took the bus at 5 p.m. and is now sitting in a university lecture-room in the city. There are sixty other people present: poor people, some leggy youths, a group of older women from an old neighbourhood, students, and a few representatives from environmental organizations in other countries, including myself. The platform creaks as Manfred Max Kneef, a bearded Chilean economist, well-respected for his ideas on environment and development inside and outside of Chile, takes the floor. In rapid Spanish he describes how the world economy has determined the recent history of the country. He points out the importance of Chilean copper, wood, fish and fruit, and notes the ascendency of multinational firms and the so-called "Chicago boys", who introduced the market economy in its most liberal form to Chile. He barely refers to the debt crisis.

After an hour, questions are fired. Ricardo is curious about the debt crisis. Max Kneef states pertinently: "The debt crisis is just a figment of our imagination. The debts are not real, they have been made up by the business community, banks and the élite. First they bleed the country dry and then they tell us we have made debts. Large amounts of capital have disappeared from the country, and huge quantities of goods have been sold for next to nothing. And now the thieves pose conditions as well, conditions which will ultimately result in more thievery! The present situation resembles one in which thieves tell us to remove the feeble locks on our doors, so that they can take the last raw materials as well." Max Kneef refers to the conditions the International Monetary Fund has imposed on Chile: that the country is no longer allowed to protect her economy, that state enterprises must be privatized. The only ones capable of buying state enterprises are foreign firms. Chile no longer belongs to the Chileans, the outside world has taken over ownership and control.

Being a Western outsider I do not feel completely at ease in this gathering. Nonetheless I ask Max Kneef what he thinks of recent efforts to realize a debt swap. At first he does not want to answer my question, but in the course of the discussion he returns to it: "Debt swaps are a good example of modern colonialism. Debts, which don't belong to us, are used to expropriate land and factories and sell them to foreign firms. A foreign firm can easily acquire Chilean currency by swapping a debt. We can't. Chileans cannot compete with these firms."

Max Kneef works for an institute which is looking for ways of changing this situation in Chile. The roads they are exploring do not lead to the industrialized countries, but take the needs of the local population as a starting point. Economic growth is not one of their basic principles.

At the end of the evening, everyone returned home. In the small community of Lampa, where Ricardo lives in a rickety house and where he grows produce in an ecological garden, the problems seem endless. A letter from Ricardo greets me when I return to the Netherlands: Lampa has been hit by a cholera epidemic, and could we help them obtain clean water?

7. Sustainable development

7.1 Development, a Trojan Horse?

Several years ago I read "The Development Game" by Leonard Frank. In this article, Frank describes the cynicism of a development worker who witnesses the umpteenth failure of a "development project". "Development, as in Third World Development, is a debauched word. Its users can't look you in the eye. Among biologists, the word means progress, the realization of an innate potential. The word is good, incontestable, a cause for celebration. In the mouths of politicians, economists and development experts like myself, it claims the same approval, but means nothing. There are no genes governing the shape of human society. No one can say of a society, as a gardener can of a flower, that it has become what it should be. It is an empty word which can be filled by any user to conceal any hidden intention, a Trojan Horse of a word. It implies that what is done to people by those more powerful than themselves is their fate, their potential, their fault."[18]

But where does the term "development" actually come from? Wolfgang Sachs wrote several essays on the archaeology of the development idea.[19]

> "Wind and snow stormed over Pennsylvania Avenue on January 20, 1949, when in his inauguration speech before Congress, President Harry Truman defined the largest part of the world as 'underdeveloped areas'. There it was, suddenly, a permanent feature of the landscape, a pivotal concept which crams the immeasurable diversity of the globe's South into a single category – underdeveloped. For the first time, the new world-view was thus announced: all the peoples of the earth were to move along the same track and aspire to only one goal – development. And the road to follow lay clearly before the President's eyes: 'Greater production is the key to prosperity and peace'. According to that yardstick, nations fall into place as stragglers and lead runners.

Using ash as fertilizer in Kenya photo: Dutch Royal Institute for the Tropics

> And 'the United States is pre-eminent among nations in the development of industrial and scientific techniques'. Clothing self-interest in generosity, Truman outlined a programme of technical assistance designed to 'relieve the suffering of these peoples' through 'industrial activities' and 'a higher standard of living'. Looking back after forty years, we recognize Truman's speech as the address inaugurating the race for the South to catch up with the North. But today we also see that the field of runners has been dispersed, some competitors having fallen by the wayside, some beginning to suspect that they are running in the wrong direction!"

Change

The term "development" shouldn't deteriorate into a play on words. While Frank explains that the term "development" is empty, Sachs explains who usually "fills it in". It is important to realize that the term is not "neutral": it is a "normative concept"; its underlying "values" must constantly be made explicit. Development implies change, and most people prefer change if this means that their situation improves. The goals of development have to be determined by our own ethics, culture and religion. But Western culture with its universal values, universal economy and universal science determines the present-day dominant pattern of development. "This concept of universalism has its foundations in the political assumptions and philosophy of the liberal European way of thought which accompanied the transition to the industrial mode of production. Political philosophers stress the importance of private initiative, private interests, private profits: competition and the principle of usefulness were its cornerstones."[20] Other cultures are based on totally different concepts of development; they have other goals and processes.

7.2 Sustainability: a common concern?

There is, however, a concern we all share. Development as it is practised today is finite. The destruction of our environment, the increase of poverty and inequity, and the exhaustibility of our resources demonstrate that our pattern of development has to change drastically. To use Pronk's words, the "non-sustainable" character of our "global village" is all too clear. It is becoming more and more obvious that the structure of the world community is violent and unjust.

In *Developed to Death, Rethinking Third World Development*[24], Ted Trainer provides us with a good overview of the contrast which has developed in our world community.

The South:

"Treatment of the peasants in Paraguay: 'The police beat them up as soon as they try to organize themselves. They have no rights at all, not even the right to protest.' The Catholic Church has been trying to get peasants to form producer cooperatives. 'One day soldiers arrived, beat up the peasants, tied them up and took them off in their trucks. Last May they destroyed the cooperative warehouse and took away the produce.' ... Twenty companies controlled by foreign capital own 30% of the land in Paraguay ... the companies controlled by foreign capital are responsible for over 80% of Paraguayan exports."
The Guardian, February 27, 1977, p. 12

"Hundreds of thousands of poor Guatemalans have been pushed off their farmlands over the last 30 years by wealthy landowners seeking high profits from export crops. Today, the richest 2% of landowning families control 80% of the country's farmland."
"Deprived of enough land to grow food, they are forced to take the only work available – two or three months a year harvesting coffee, cotton and sugar cane to be shipped to well-fed people thousands of miles away. Their pay: $2 a day or less. No wonder the poorest 50% of Guatemalans – mainly Indians – get only a little more than half the calories that they need. No wonder hunger-related diseases kill *half* of their children before they reach the age of five."
"Guatemala: hungry for change", *Food First Action Alert,* March 1983

"The richest 1% of Haitians monopolize 44% of the nation's income. As for the other 5 million Haitians, the World Bank estimates that 80% survive on about $100 a year. Some of those lucky enough to have jobs work for the more than 250 US companies assembling baseballs, electronics and clothing – all for export. The usual wage is $2.64 a day (33 cents an hour) and unions are banned. Large landowners control the best land, mainly used for export

crops, including sugar, cotton and coffee. Poor Haitians have been forced onto marginal lands on mountainous slopes."
"Haiti", *Food First News*, Summer 1983

"Peasant farmers in Thailand sometimes have to pay interest of more than 100% per month on loans. Trade unions and strikes are outlawed."
P. de Beer: Thailand's chaotic economy, *The Guardian*, August 28, 1977.

"In many countries, tenants have to hand over to the landlord 50-60 percent of their crop as rent ..."
S. George: *How the Other Half Dies*, 1986

The North:

"Americans put more fertilizer on their lawns and tennis courts than India uses for all purposes."
New York Times, June 14, 1979

"In 1975, Americans spent $4,300 million on barber shop, beauty and bath services."
P. Harrison, *Inside the Third World*, 1979, p. 361

"Australians drink on an average 65 litres of soft drinks every year."
National Times, September 14-20, 1980

"Australians own 4,139,000 dogs; $180 million is spent on pet food every year."
Sydney Morning Herald, Febrary 25, 1980

"The Japanese sent 22 billion New Year cards at the end of 1978."
Sydney Morning Herald, January 5, 1979

"Australians have $100 million worth of air conditioning in their cars."
National Times, December 16, 1978

"In the 1950s, Americans spent $5,339 million every year just to change their car models."
F. Fisher *et al.* The costs of automobile model changes since 1949, *Journal of Political Economy*, October 1962

"In 1986, Australia imported $152 million worth of beverages, $1,154

million worth of cars, and $84 million worth of caviar."
Sydney Morning Herald, May 16, 1986

"There are between 12 and 20 electronic motors in the average American kitchen."
A.B.C. Radio, *Technology Report,* April 8, 1987

"America produced, and threw away, 88,000,000,000 tin cans in 1981."
U.S. Bureau of Mines, June 1983

7.3 Homo economicus: the materialistic identity

"It was not until after the Second World War, precisely in the age of 'development', that the Third World countries moved into focus under this conception of the world; they were perceived for the first time from a material-centred viewpoint. Spurred on by the experience of societies investing all their physical and mental energies in the propagation of things, development strategists perused the world and, lo and behold, discovered an appalling lack of useful objects wherever they looked. However, what was of primary importance in many villages and communities – the tissue of relationships with neighbours, ancestors and gods – more or less melted into thin air under their gaze. For this reason, the popular image of the Third World was dominated by have-nots desperately battling for mere survival; but, whatever constituted their strength, honour or hope remained out of sight. Although such a definition fails to capture the realities of the lives of many people, it still provided the basis for the formulation of programmes of global goodwill. A classic example of this occured when John F. Kennedy called upon Congress in March 1961 to finance the 'Alliance for Progress': 'Throughout Latin America millions of people are struggling to free themselves from the bonds of poverty, hunger and ignorance'. In the wake of such an interpretation of the aspirations of people all over Latin America – from traders on the Gulf of Mexico to cattle farmers of the pampas – the strategic conclusion was self-evident: 'To the North and the East they see the abundance which modern science can bring. They now have the tools of progress within their reach.'"[22]

Herman Daly and John B. Cobb write: "What is excluded from

Homo Economicus are concerns for other people's satisfactions or sufferings that do not express themselves as one's market activity. For example, *Homo Economicus* takes no pleasure when a neighbour receives a gift from someone else or a promotion, or when a philanthropist endows a park for underprivileged children. This adds to the utility function of the philanthropist, but not to the other members of the community. Similarly, *Homo Economicus* is not envious of the neighbour's new car or pained by defeat in competition for an honour. *Homo Economicus* knows neither benevolence nor malevolence in any of these instances, only indifference.

"When *Homo Economicus* is compared in these respects with flesh and blood people, the contrast is striking. *Homo Economicus* is indifferent to relative position in society, but in the real world much of the satisfaction in life felt by individuals is a function of how they stand in relation to other individuals; in other words, their relative status in their community. On the whole those who are relatively better off report that they are happier than do other members of their society who are relatively worse off. This is what most of us would expect. But when comparisons are made over time or between societies, there is little difference in self-reported happiness from one society to another. That is, the fact that at a later date most people are consuming much more absolutely does not have much effect on happiness. And people in affluent societies do not on the whole report that they are happier than those in poorer ones. In short, the absolute level within a given society contributes considerably. An abstraction that neglects this points entirely cannot provide adequate guidance for policy."[23]

A world cut to measure

Poverty is a relative concept: we Northerners measure it in terms of material prosperity. Are people with more furniture, more implements in the kitchen, a more expensive car or a larger house happier than others? It has never been proven that the measure of happiness is wealth. Twenty years ago people were not "unhappier" than they are now, and we do not expect our grandchildren to be so much more happy than we are. Happiness and social inequality are, however, linked. The larger the differences in society, the greater the urge to grow and possess more. Inequality makes people unhappy and stimulates growth. Inequality is the root cause of the

present environmental crisis and the mainspring of war; it encourages greater disparities, poverty and underdevelopment.
This would suggest that equality is the answer: equal opportunities, one world, one global community with social facilities for every person. Jan Pronk seems to strive for this goal in *A World of Difference*. But equality is a non-existent concept. If we try to attain it we will invariably end up with a world based on Western values, in which economic growth and waste go hand in hand. The funny thing about "inequality" is that the way it is defined determines the way it is used as a measure. Let's take a look at the underlying values of the concept and the parties that define them.

There are many other "immaterial values", such as friendship, culture, knowledge and the quality of our individual surroundings. They do not fit into our economic thought. We talk about the car, the washing machine, and the level of income as if that is equality. This measure of comparison will never contribute to realizing the goal of equality. The international division of labour and the market – constrained by the failure of its ideal, free trade – irrevocably lead to material inequality. The compulsion to grow is intrinsic to the system. As distances become smaller – because of aeroplanes, satellites, telephone connections and television images – people will become increasingly more unhappy. They will want to speed up their rate of growth and they will be prepared to use the very last piece of fertile land on this planet to become a little richer than their neighbour.

To stop the compulsion to grow, we will have to upgrade "other values". The problem is illustrated by the devout man in India who had been praying to God for many years. He became so very holy that God finally revealed himself and said solemnly: "Speak, and your wish will be fulfilled." The praying saint answered: "Then I would like to ask you to step aside a little, as you are standing in the sun I have been enjoying for many years now." In the North we are trying, in our own clumsy way, to introduce non-economic values in our decision-making processes. The environmental movement demands that "external factors", such as pollution, be included in all price calculations. These are eco- or green taxes, which may be effective at a certain level. But changes will only be achieved when the compulsion to grow goes.

We have to say "enough is enough" and set definite limitations to growth. We will have to formulate stringent norms and goals. Reduce car traffic by half! Ration petrol! Create values that count!

The open market dominates

Development, defined as material growth, has become the doctrine of modern colonialism. As the problems associated with this doctrine become more urgent – the difference between rich and poor is sky-rocketing, violence is on the rise and the environment faces destruction – the doctrine itself is being questioned.

Jan Pronk describes two central dilemmas:

> "Looking back over forty years of development and against the background of scenarios in which a number of previously neglected factors appear to be coming to the fore – the possibility of not only relative but absolute physical scarcity due to an ecological crisis; the globalization of the market, making a distinction between internal and external factors almost impossible – it is very questionable whether choices can be avoided. Is economic growth still possible? The conclusion drawn towards the end of the debate on development theory was that, while it is a precondition for development, it is not enough in itself. Is this still true? Does continuous growth not harm the natural basis for any form of human development?
>
> "And – the second dilemma – whom does the development process actually benefit? Is there not, in fact, far more at stake than better distribution of the benefits of the process? Is autonomy not the goal? Autonomy for individual people in a society, so that they have the opportunity from the outset to decide for themselves what development should mean, what life and existence should signify? If that is so, the question of power arises, a choice has to be made after all, development and the achievement of human rights become identical." (Pronk, p. 38)

Pronk, one of the leading politicians in the international development arena, questions economic growth and the increasing dependence on the international market. Most political leaders and economic analysts believe that economic growth (meaning growth of material consumption and production) is essential. This sort of economic growth can only be achieved through an "open market". The market's "invisible hand", to use Adam Smith's metaphor, leads decision making in a particular direction. Many have cursed the open market for this reason. Others argue that the "market" is a given, unavoidable fact. They believe that people should fight for their rightful position in the market. A major section of the Third World movement in the North thinks this way.

It is more important to research the advantages and disadvantages of free economic enterprise than to discuss what we think of the open market. Daly and Cobb write:

> "There are some important qualifications and limitations to the virtues of the market, but before considering them let us put aside a confused objection. Some people have simultaneously condemned both the market, with its reliance on profit, and the concentrated power of centralized bureaucracy. This involves an inconsistency. If one rejects the very concept of profit then everyone must be a salaried employee – no independent artisans, professionals, small farmers, merchants, and so on. If all are employees, then who is left over to be the employer? Presumably the State – a single, massive, centralized bureacracy. If one favours independence, participation, decentralized decision making, and small- or human-scale enterprises, then one has to accept the category of profits as a legitimate and necessary source of income. There is plenty of room to complain about monopoly profits, but that is a complaint against monopoly, not against profit *per se*. People who want to abolish the category of profit and at the same time foster decentralized decision making and participation in economic life have not thought very clearly about their proposal. If one dislikes centralized bureaucratic decision making then one must accept the market and the profit motive, if not as a positive good then as the lesser of two evils. A third alternative (as opposed to varying mixes of two basic ones) has not been identified."[24]

What we witness in our modern world is that economic action on the open market tends to destroy its own freedom. Daly and Cobb once again:

> "Three broad categories of problems with the market have been identified by economists: (1) the tendency for competition to be self-eliminating, (2) the corrosiveness of self-interest on the moral context of community that is presupposed by the market, and (3) the existence of public goods and externalities."[25]

The disappearance of political and geographical boundaries, a product of the technological revolution, and the extreme use of fossil fuels are the most important forces behind this "self-destruction".

7.4 The real contradiction

The North consumes about three-quarters of the world's resources, but is home to only a quarter of its population. An average person in the industrialized world uses eighty times as much energy as an inhabitant of Sub-Saharan Africa. The population of the United States has used more energy in the past fifty years than humanity has burned up in its entire history. If everyone were to consume as many resources as the average American citizen, the world's annual production would have to be 130 times higher than it was in 1979!

The Industrial Revolution was possible because of the transfer of raw materials from the South. Without the slaves, gold, silver, spices and other resources from the colonies, we would never have been able to amass sufficient capital and raw materials to carry the revolution through. The term "transfer" is actually too mild, it doesn't convey the human impact of the colonial process which caused the death of millions of Africans, Indians in South and North America, Polynesians in the South Pacific and Aboriginals in Australia. Even now people die in the course of the "transfer" of coffee, cocoa and ore.
The international division of labour, in which the South supplies the raw materials which the North transforms into manufactured products, generates a large number of problems in the North and in the South. The greatest problem in the North is "waste". Industrialized countries have so many imports, they do not know what to do with the by-products. The large quantities of manure which are strewn on the fields in the Netherlands cause soil destruction and acid rain. Chemical wastes are dumped on land or in the sea – or they are exported to the Third World or Eastern Europe.
The South has completely different problems. It also has a waste problem – often caused by Western firms – but the main issues are related to the export of raw materials. The local population no longer has access to its own resources, because these are taken exclusively for the foreign market. Natural resources are becoming exhausted. Desertification, deforestation and hunger are the results.

The exploitation of the South by the North has three components. The first is the structure of the international economy: the sale of resources and labour power, the debt problem and the role of banks and the Structural Adjustment Policies. Once countries are tied up in the exploitative system, their economies are dictated by investors and creditors. It is difficult to find a way out. International agreements (such as the General Agreement on Tariffs and Trade, GATT) establish rules which permit increasingly larger volumes of raw materials to be transferred to the North.

Secondly, the North has to ensure the continuation of the exploitative relationship. Exploitation is not something people agree to undergo. It is regulated by force. The state has the necessary means – army and police, to do this. In general, however, there is no need to use physical violence, as the social structure itself ensures compliance. Unions can be prohibited, social relationships destroyed, people compelled to move. Without this structural violence, exploitation would never be what it is today.

The third component is the ideological subordination which can be called "modernization" or "Western domination". Education, thinking, the relationship of people to their surroundings: everything is being Westernized. Every corner of the earth is becoming integrated in this international society. The local identity, the relationship with the world around them, the social order, and the thousand-year-old histories unique to every culture and way of life, often guaranteed a precarious balance with the environment. The church played an important role in the destruction of original cultures in the early colonial period.

Rich versus poor, exploiter versus exploited, these are the central relations in this story. I present them as a North-South scheme, but similar relations can be found in any society. Local élites in the Third World often have interests very similar to those of the North.

Table 7.1 Distribution of World Consumption, averages for 1980-1982

		Developed countries		**Developing countries**	
Commodity	*Units of per capita consumption*	*Share in world consumption (per cent)*	*Per capita*	*Share in world consumption (per cent)*	*Per capita*
Food					
Calories	Kcal/day	34	3,395	66	2,389
protein	gms/day	38	99	62	58
fat	gms/day	53	127	47	40
Paper	kg/year	85	123	15	8
Steel	kg/year	79	455	21	43
Other metals	kg/year	86	26	14	2
Commercial energy	mtce/year	80	5,8	20	0,5

Source: WCED estimates based on country-level data from FAO, UN Statistical Office, Unctad and American Metal Association.
Table 1.2 from the Brundtland Report.

7.4.1 Three central antitheses

The debate on environment and development is fashionable. Everyone seems to be in favour of sustainable development. The term is used to mean various things; it is misused to promote a development model or certain products or processes as being ecologically and socially "sound". The serious participant in the debate on sustainable development will have to take a stand on three essential antitheses: growth versus equilibrium, de-linking versus integration, and participation versus economic domination.

Growth versus equilibrium

Many people consider the real foundation of development to be continuing economic growth, and believe that growth can finance investments in the environment. The Brundtland Report champions this point of view.

You can tackle part of the environmental problem in this fashion. For the large-scale, and particularly the global, environmental problems, the relationship between the world economy and the biosphere is important. It has been calculated that to stabilize the present atmospheric concentration of carbon dioxide, global emissions will have to be reduced by at least sixty percent. If a highly industrialized country like the Netherlands were to maintain its present level of production and employ the best technical alternatives, a reduction of thirty percent could be achieved. The rest would depend on measures to reduce the volume of production.

The term "economic growth" is no longer sacrosanct in Third World countries. Gross National Product (GNP) is the usual indicator of economic progress, but it cannot be used to measure growth. The "informal sector" – the economic activities which are not registered – is not included in the GNP. The destruction of nature and the environment, which is actually negative production, is also not measured in the GNP.

In the future, the issue of enterprise will be far more important than the issue of growth. To achieve sustainable development, some sectors will have to shrink, others will have to grow. Labour-intensive sectors, such as biological agriculture, and small-scale industry, will have to grow. Capital-intensive sectors, such as the mining and chemical industries, will have to shrink.

De-linking versus integration

Where can the growing world population find employment if not in labour-intensive activities? It is not in the interests of labour-intensive economic sectors to integrate in the world economy. Large firms are generally better qualified to compete internationally. GATT trade agreements aim at increasing economic growth by liberalizing international trade and abolishing protectionist policies which defend the country's own market. Similarly, the trend in debt repayment negotiations is to compel Third World countries to integrate their economies in the world market, and to join a game in which the rules are decided by others. As boundaries between East and West and South and North fade, Western values and lifestyles gain ground.

ere is only one way out of the present crisis: to move towards a society which operates on a human and a natural scale. The starting point will have to be respect for the many cultures of this planet, cultures which have originated and developed over a long period of time in the framework of the local natural environment. This local evolution must continue. It must not be destroyed in the violence of a Western development model which eradicates all positive feedback mechanisms, whether they be social, cultural, religious, economic or environmental.

It is not appropriate to idealize "traditional cultures". At the same time we have to realize that a sustainable society can only be built on the foundations of its "own" culture and tradition. Development will not be imposed from the outside, but will grow because of an internal dynamic.

This is illustrated by the crisis in agriculture. There are many serious and widespread doubts about concepts such as the agricultural Green Revolution, while low input "Sustainable Agriculture" enjoys a great deal of attention. Alternative concepts of this kind do not function (yet) at the level of the world market; they are strong on the level of the local market. Contrary to popular opinion, "de-linking" does not lead to absolute economic self-sufficiency. It contributes to the distribution of power and to the reinforcement of local styles of government and administration. The people themselves can determine their own development.

Participation versus economic domination

Like "sustainable development", the term "participation" is often bandied about. Participation is more than just the contribution of a local population to an external initiative. It begins with the historical and cultural identity of the inhabitants. By participating, the local population becomes involved in the preparatory activities, in the development, research, maintenance, extension and evaluation of development projects. The population has its own responsibilities and is able to take decisions autonomously.

Non-governmental organizations have an important part to play in this participation. All over the world people are organizing themselves: in unions, women's groups, environmental groups. These groups often play an important role in the process of change. Their networks reach down to the level of the individual, and they have the capacity to translate people's wishes into policy. In addition, they can exert political influence at various decision-making levels.

Inevitably, the problem of unequal power has to be dealt with on the road to empowerment and autonomy. Twenty percent of the world population would not be able to command control over seventy-five percent of the available resources if they did not have economic clout and instruments of power at their disposal. This is why land reforms, democratization and devolution of power to local and regional producers and consumers are so important. The creation of fundamentally different power relationships is essential to "sustainable development".

Countries have to maintain their right to protect their own market and culture.

7.4.2 Unequal competition

There is nothing wrong with competition, but if the participants are fundamentally unequal, competition systematically eliminates the weaker party. The inequality between nations has dominated international relations since the colonial period. Countries with raw materials supply Northern firms with cheap inputs which are transformed into consumer goods. The Northern firms skim off the profits and shape the market. Every market can be reached, thanks to the development of international transport and communications – the international division of labour has its repercussions in all corners of the globe. How can the business of a small-scale shoe-maker in Bombay, Seoul, or an isolated valley in the Himalayan mountains of Nepal, compete with plastic shoes which are dumped on the market from all sides, for ridiculously low prices, heralded by a sophisticated advertising campaign underlining the modernity of the product? That is not competition, that is execution! Economic domination marginalizes large sectors of the world community: participation is made next to impossible because of economic domination.

7.4.3 Losing the market place

The open market can also lead to the disappearance of original, and often very complex, marketing structures. "The exchange of commodities during fairs and market days is a practice of ancient origin. The market in this sense existed in the time of feudalism, and continues to exist even in communist societies."[25] These markets

formed, and in many places still form, the nucleus of communities. Each market had its own identity, culture and history. Under colonialism the importance of these regional markets has been devalued. The regional markets have been replaced by a market in which everything has a commodity value. "Labelling things as 'resources' takes away whatever protective identity they may have and opens them to intervention from the outside. Looking at water, soil, animals, people in terms of resources reconstitutes them as objects for management by planners and for pricing by economists. Even if they are renamed 'resources' in order to maximize their efficient use, because of the cultural fall-out from the all-embracing economic cloud, it will, in the future, be much more difficult to have any intrinsic respect for them."[26]
The fading boundaries, the reinforced international relationships, and the division of labour undermine the identity of a community. Objects with complex values are reduced to being commodities. The most tragic example of this process is the privatization and commercialization of land ownership. The ground was given a price tag and became a saleable commodity. Yet it was the basis of life for the community which lived on it. Destruction of communities all over the world was preceded by the privatization of their land. The same happened with labour, which was made "an exchangeable commodity to be valued and exchanged according to supply and demand ... Capital was abstracted out of the social inheritance, no longer to be treated as a collective patrimony or heirloom, but as an exchangeable source of unearned income to individuals."[27]

Sachs summarizes the main elements of the conflict in a critique of reports like *The State of the World:*

> "Certainly, interpreting the state of the world chiefly in terms of 'resources', 'management' and efficiency may appeal to planners and economists. But it continues to promote development as a cultural mission and to shape the world in the image of the West. The reports do more than simply propose new strategies; they also tell people how to see nature, society and their own actions. The more their language is adopted around the globe, the more difficult it will be to see nature in terms of respect and not as a resource, society in terms of the common good and not of production, and actions in terms of virtue and not of

> efficiency. To put it in a nutshell: they promote the sustainability of nature and erode the sustainability of cultures. And this, for sure, will not benefit nature either."[28]

"I sympathize, therefore, with those who would minimize, rather than with those who would maximize, economic entanglement between nations. Ideas, knowledge, art, hospitality, travel – these are things which should of their nature be international. But let goods be homespun whenever it is reasonable and conveniently possible; and above all, let finance be primarily national." J.M. Keynes

The international division of labour is the basis of the modern economy. The principle of division is very simple: one person is good at typing, the second at driving a truck, and the third person is a capable manager. Each has his or her "comparative advantage", as it is called by the disciples of international trade, and together they have the potential to start up a transport business. This principle can be applied at a national level: one firm does the administration, another takes care of production, and a third firm transports the goods; on a global level the South delivers cheap labour and raw materials which the North processes into the final product it trades. It seems so logical. The countries of the South have ample labour power and raw materials but lack the knowledge to process and trade them. Why shouldn't they make use of their comparative advantage and leave the other activities to other countries?

But what happens if capital choses its own path? The party with the most power plays off the suppliers of raw materials and cheap labour and keeps prices low by prohibiting trade unions. Enterprises have become very mobile and can easily move from one place to another. A factory in the Philippines can be transferred to South Korea or Mexico with a minimum of hindrance. The South does not seem to have a choice: the North invests its capital where the returns are highest, and transfers profits and cheap products back to the parent companies. In 1923 the famous economist Keynes considered free trade "not only as an economic doctrine which a rational and instructed person could not doubt, but almost as a part of the moral law." Ten years later he stated:

> "It does not now seem obvious that a great concentration of national

> effort on the capture of foreign trade, that the penetration of a country's economic structure by the resources and the influence of foreign capitalists, that a close dependence of our own economic life on the fluctuating policies of foreign countries, are safeguards and assurances of international peace. It is easier, in the light of experience and foresight, to argue quite the contrary ... Advisable domestic policies might be easier to compass, if, for example, the phenomenon known as 'flight capital' could be ruled out. The divorce between ownership and the real responsibility of management is serious within a country when, as a result of joint-stock enterprise, ownership is broken up between innumerable individuals who buy their interest today and sell it tomorrow and lack altogether both knowledge and responsibilty towards what they momentarily own. But when this same principle is applied internationally, it is, in times of stress, intolerable – I am irresponsible towards what I own and those who operate what I own are irresponsible towards me."[29]

Ted Trainer argues:

> "No doubt there is a place for some amount of trading within a satisfactory framework of development. However, Third World countries easily find themselves drawn into far too much trade and other economic interactions with the developed countries and their corporations. It is very much in the interests of the transnational corporations and banks of the rich countries that there should be more rather than less trade. The more international business that takes place the better it is for them, and it is not surprising that conventional development theory encourages as much increase in world trade as is possible."[30]

There are very few examples in which international trade contributes to a real improvement in a community's situation without at the same time causing disastrous consequences somewhere else. This fact is indigestible for many Third World activists, who advocate that Third World countries should win a stronger position in the international market. This conflict of de-linking (autonomy) versus integration will have to be solved before constructive change can begin.

7.4.4 Forgotten factors

The price of a product reflects only part of its costs to society.

External costs, such as illness due to unhealthy labour conditions, or pollution, erosion and desertification, are seldom part of price calculations. A popular demand of the environmental movement is to include these costs in prices, to automatically induce consumer patterns to shift to products which are less polluting, destructive and anti-social. The discussion of "green taxes" is beginning to reach policy- makers as well, and it will not be long before an extra "environmental tax" is imposed on the use of energy.

The idea is not, however, as simple as it seems. The environmental movement wants to levy "regulating taxes", which incorporate the external costs of the product (what a term!) and induce consumers to choose products with low total costs. This approach will not fundamentally reform the market but will correct it through market instruments. Studies point out, however, that price increases will have to be very substantial – fifty to one hundred percent – to achieve results. Minor increases will have little effect. The effect of only partial inclusion of external costs in calculations may be negative. Environmentally friendly products are often more expensive. As long as the players in the economic theatre are the same and their relative power positions do not change, a price increase will create an extra push on production and consumption: the loss in capacity to consume (a kind of environmentally induced inflation) will create pressure to increase the incomes of the consumers. Within the geographically dominant regions (the Centre, the North) this will create an increased need for productive employment and markets. At the same time, it will create an extra money-flow – taxes managed by institutions like the national and supranational state (or World Bank) which is dominated by the powers within the Centre (since they own the productive capacities and hold the positions of power on the national and international level). The extra money-flow will be used for investments in the interests of the Centre, which will often not be in the interests of the periphery or of the local and household levels of society. The investments will be used to increase the efficiency of production and transport in order to create a new competition within the price structure. Ecological modernization is already forcing smaller enterprises off the market. It is also imaginable that, with prices of energy-intensive products (or polluting products) rising drastically, a real shift in priorities will take place. This would be a drastic way of

preventing the problems described above. This sort of measure will have to accommodate the complexity of the capital market and the flexibility (including the geographical) of modern enterprises.
It is senseless to discuss "economic growth" and not define its parameters. An equilibrium in the capacity of our biosphere is a prerequisite if life on this planet is to be sustained. At the same time, economic activities have to be created which will give people the scope to emancipate themselves, to develop, to build up self-reliance and local control over resources and power, and which will aid the participation of marginalized peoples. The conflict between growth and equilibrium is the conflict between colonies and the colonizers, between the periphery and the Centre which has defined the parameters and the targets.

7.5 Shifting frames of reference

Choices have to be made if we are to resolve the dilemmas described in the three central antitheses. This book does not provide a blueprint of correct choices. It presents an analysis of perspectives offered by environmentalists and feminists, and it links these perspectives to grass-roots experiences. It leads us to the concept of "autonomy". Communities cannot afford to wait until the outside world solves their problems. The outside world, international and national authorities, should not have the power to solve a community's problems. The power from within the communities themselves, and not economies, should grow. The role of the state should be to protect the interests of the communities, safeguarding the market-place from the interests of the establishment.

7.5.1 World view in pioneer and climax situations

"Pioneer situation" and "climax situation" are concepts used in the environmental sciences. A tidal area, a building site, or a forest which was recently chopped down, all provide pioneer situations in which "pioneers" (generally plants and insects) can multiply rapidly. They proliferate like a plague, covering the naked soil with the first plants and the first ecosystem. You can see the same thing happening in your

own garden. If you weed the garden before you go away for a holiday, it will be full of weeds, often only five or six varieties, on your return. These are the pioneer plants, and they generally flower before you have time to get rid of them. They are temporary even if they are not weeded they will disappear with time. If the pioneers are left alone, other plants will make use of the protection from the sun and the wind which the pioneers provide. A few years later the area will be covered with all sorts of vegetation; after many years it will have become a small forest. If it were to remain untouched for a century or more, but this is unlikely, it might become an oak forest. By this time the number of plant varieties will have stabilized, and the ecosystem will have reached its peak, its climax.
The pioneer situation is very productive: a single species grows very rapidly. Corn fields, potato fields or other mono-cultures demonstrate this. Within a few months the plants are mature, bear fruit and can be harvested. The soil remains bare, waiting for new pioneers.

The productiveness of a pioneer system is very much dependent on external inputs. During the first cycle of rapid production the system becomes exhausted, and chemical fertilizers, irrigation and pesticides have to be applied in subsequent cycles to maintain production levels. In the long term, when the economic, environmental and social costs of investments are taken into account, the total production is relatively low.

The Green Revolution experience in the Punjab in India demonstrates that after a period of intensive agriculture lasting ten to twenty years, production declines to a level which is too low to feed all mouths. Investment levels in fertilizers, pesticides and irrigation are so high that a dramatic decrease in production is unavoidable. As a result, the Punjabi farmers who cannot afford the necessary investments leave the area to find a new livelihood elsewhere. Many of them end up in landless misery. These are the environmental refugees, who swell the populations of the ever-burgeoning cities.

In view of their long-term productivity and external investments, tropical rain forests, mangrove forests and coral reefs are typical "climax systems", capable of maintaining a higher level of production. The system can maintain itself by reducing external losses. Production cannot simply be calculated in terms of market value: a tropical forest provides a lot more than wood alone. It provides an agrarian society with "certainty in the supply of food by providing basic food and, more often, extra nutritional substances, protection from shortages, the ingredients of traditional medicines, the material for building houses, and instruments for other economic activities such as farming, fishing, hunting and, finally, income and work – collecting, trading and producing products."[31]

Internally, a climax system can be very dynamic. A tree dies, fungi and insects live on the dying and dead wood, and the tree soon decays completely. But nothing has been lost. The old biomass has been transformed, very swiftly and very efficiently, into new living matter; the cycle is completed.
In contrast to the Western development model, the Gaia Hypothesis views the earth as a climax or mature ecosystem: high productivity is not compatible with the invulnerability we have learned to count on.

7.5.2 Women

The feminist movement has given momentum to our thinking about the relationship between people and nature. According to radical feminism, the oppression of women is rooted in patriarchy, the systematic domination by men which fixes the position of women to their reproductive biological functions as mothers and sex objects. Radical feminism takes the critique of the current understanding of the world several steps further than liberal feminism or traditional marxist feminism. These liberal and marxist feminist theories and proposals for policy changes emphasize the lack of equal rights in the public domain and the need to change society to bring about equality between women and men. Radical feminism does not necessarily lead to the romanticist view which portrays woman as the descendents of mother nature, charged with the care of "her" creation. Many feminists repudiate the women/nature connection and condemn any links with ecology, because the association of women and nature strengthens the oppressive stereotypes and perpetuates subordination. Cultural radical feminists accept the association of women with nature.

Erosion in the Himalayas photo: Wallrafen/HH

They condemn the male culture for its agressiveness, individualism, and hierarchical value thinking.
Susan Griffin vividly describes the impact of patriarchy in her book *Women and Nature*[32].

> "He breaks the wilderness. He clears the land of trees, bush, weed. The land is brought under his control; he has turned waste into a garden. Into her soil he places his plow. He labors. He plants. He sows. By the sweat of his brow, he makes her yield. She opens her broad lap to him. She smiles on him. She prepares him a feast. She gives up her treasures to him. She makes him grow rich. She yields. She conceives. Her lap is fertile. Out of her dark interior, life arises. What she does to his seed is a mystery to him. He counts her yielding as a miracle. He sees her working as effortless. Whatever she brings forth he calls his own. He has made her conceive. His land is a mother. She smiles on the joys of

her children. She feeds him generously. Again and again, in his hunger, he returns to her. Again and again she gives to him. She is his mother. Her powers are a mystery to him. Silently she works miracles for him. Yet, just as silently, she withholds from him. Without reason, she refuses to yield. She is fickle. She dries up. She is bitter. She scorns him. He is determined he will master her. He will make her produce at will. He will devise ways to plant what he wants in her, to make her yield more to him."

After describing the violent rape of mother earth, Susan Griffin goes on to describe the development of science and our modern knowledge system:

"He deciphers the secrets of the soil. (He knows why she brings forth.) He recites the story of the carbon cycle. (He masters the properties of chlorophyll.) He recites the story of the nitrogen cycle. (He brings nitrogen out of the air.) He determines the composition of the soil. (Over and over he can plant the same plot of land with the same crop.) He says that the soil is a lifeless place of storage, he says that the soil is what is tilled by farmers. He says that the land need no longer be fallow. That what went on in her quietude is no longer a secret, that the ways of the land can be managed. That the farmer can ask whatever he wishes of the land. (He replaces the fungi, bacteria, earthworms, insects, decay.) He names all that is necessary, nitrogen, phosphorus, potassium, and these he says he can make. He increases the weight of kernels of barley with potash; he makes a more mealy potato with muriate of potash; he makes the colour of cabbage bright green with nitrate; he makes onions which live longer with phosphates; he makes the cauliflower head early by withholding nitrogen.

"His powers continue to grow. Phosphoric acid, nitrogen fertilizers, ammonium sulphate, white phosphate, potash, iron sulphate, nitrate of soda, superphosphate, calcium cynanamide, calcium oxide, calcium magnesium, zinc sulphate, phenobarbital, amphetamine, magnesium, oestrogen, copper sulphate, meprobamate, thalidomide, benzethonium chloride, Valium, hexachlorophene, diethylstilbestrol.

"What devices she can use to continue she does. She says that the pain is unbearable. Give me something, she says. What he gives her she takes into herself without asking why. She says now that the edges of what she sees are blurred. The edges of what she sees, and what she wants, and what she is saying, are blurred. Give me something, she says. What he gives her she takes without asking. She says that the first

> pain is gone, or that she cannot remember it, or that she cannot remember why this began, or what she was like before, or if she will survive without what he gives her to take, but that she does not know, or cannot remember, why she continues.
>
> "He says she cannot continue without him. He says she must have what he gives her. He says also that he protects her from predators. That he gives her dichlorodiphenyltrichloroethane, dieldrin, chlorinated naphthalenes, chlordan, parathion, Malathion, selenium, pentachlorophenol, arsenic, sodium arsenite, amitrole. That he has rid her of pests, he says.
>
> "And he had devised ways to separate himself from her. He sends machines to do his labor. His working has become as effortless as hers. He accomplishes days of labor with a small motion of his hand. His efforts are more astonishing than hers. No longer praying, no longer imploring, he pronounces words from a distance and his orders are carried out. Even with his back turned to her she yields to him. And in his mind he develops the means to supplant her miracles with his own. In his mind, he no longer relies on her. What he possesses, he says, is his to use and to abandon."

The relationship between women and development is now being addressed by all the important development institutions. There is hardly a multilateral bank or development organization that lacks a "Women's Department". Most institutions strive to increase the scope of development efforts to include women and to organize "gender-training". But, though important, this does not constitute a real change within the institutions. Real changes will only happen when the knowledge system as such changes and the institutions open up to a dialogue with people at the grass-roots.

Autonomy

The principal complaint of the women in Sunita's new colony, part of the Mahaweli development project in Sri Lanka, was their loneliness and social isolation. "We are going crazy, there is no one to talk to." The women, who are mostly young, depend primarily on their husbands and young children for company. In the beginning, Sunita enjoyed being no longer subjected to the control of her older relatives. But the control exercised by husbands made life tougher than ever. Parents, brothers and sisters could no longer be called upon in the event of marital conflicts. There had been many conflicts in the colony. Most were hardly worth mentioning, but people were so stressed that a small incident could easily become a violent conflict. The men decided to avoid continuing stress by categorically forbidding their wives and daughters to have any contact with other women. They feared the escalation that in their minds would be inevitable due to the "naturally quarrelsome and emotional character of women". They frequently used physical violence to enforce this resolve. Most of the women had been beaten up more than once by their husbands. The women developed a survival strategy, internalizing the ideology of women's bad character. Joke Schrijvers, a Dutch researcher who lived in Sri Lankan villages for some time, described an incident:

> "When we went to visit her again, Bisomenika flew out at us: she had terrible problems with her husband. He had changed completely since they had come to live in the new village. He used to be a good husband but now he was away more and more often, and when he returned he ill-treated her. 'Look, I've lost all my front teeth, that happened soon after we came here.' Once he also broke her upper right arm, she said. Before coming here, he never hit her. Bisomenika explained his altered behaviour by blaming another woman: she had soon noted that her husband was going with another woman in the village. He tried to conceal it, but she knew. He went to that house over there, to that whore. That is a women who can snare all the men. Everyone knows that, because at first she was the only woman in the village when only men were here to build the houses. That's why. That woman also gave her services to all the officials from the project. She had many things in her house, bags of rice, a sideboard and other fine furniture (Bisomenika gave a fairly complete description of the contents of the other woman's house). That whore had so many things because she provided her services on credit. After the harvest she was given rice, half a sack or a whole sack according to the value of her services. Or money. According to Bisomenika, most of the married couples were at odds because that woman knew how to lure the men.
>
> "Once, six months ago, she had fought with the woman on the path to the river. They had pulled each other's hair out, but a man came immediately to separate them. The women in the nearby houses had stood watching. After that her husband had forbidden her ever to visit other women or to talk to them on the river path."[34]

It is difficult to imagine a way out of this sort of violent situation. To add a woman's desk to the company responsible for the Mahaweli irrigation project would not really help, although in itself it would be a good thing. It is more important to assess whether the entire project serves the interests of women. Some feminists use the term "autonomy" to express the terms of reference for activities which are in the interests of women, and therefore in the interests of society as a whole:

1 women's control of their own sexuality and fertility; forms of shared mothering, between women or between women and men;
2 a division of labour which allows women and men equal access to, and control over, the means of production;
3 forms of cooperation and organization of women which will enable them to control their own affairs;
4 positive conceptions of gender which legitimate women's sense of dignity and self-respect, and their right to self-determination.

Jan Pronk states in *A World of Difference:*

> "If these four elements are to be achieved, current ideologies, which deny such concepts, will, of course, have to be transformed to produce a development ideology which is based on the principle of autonomy. In the case of physical autonomy, for example, this means the right of women to decide for themselves with whom they enter a sexual relationship and the right to choose whether or not to have children. In other words, freedom from the pressure of endless childbearing and freedom of choice of means of family planning. This will not come about of its own accord. Consciousness-raising, and cooperation between and organization by women are essential prerequisites. Development aid can be used to support such cooperation." (Pronk, p. 205)

Jan Pronk extends this conclusion, using the term autonomy in a more general sense:

> "The four elements listed above cannot, however, be considered or used in isolation from one another, as was sometimes the case with development models in the past. They have a meaning only in relation to one another. Autonomy may be used as an analytical concept to shed light on this relationship. When planning and formulating development activities – programmes and projects – it is also necessary

Poster at the 1985 international forum for women, held in Nairobi, Kenya
photo: Bijeen

> at an early stage to gain insight into and information about power relations, in order to create opportunities for representation and self-determination. Dutch development policy will be designed in such a way as to satisfy this condition." (Pronk, p. 205)

Self-determination as the starting point of a development strategy. Not as a simple Western answer expressed as "equality" or "freedom", but within the specific cultural, social, historical, and religious context of the hundreds of thousands of different communities around the world. Understanding the conflicts in Sunita's new colony helps us to get a basic understanding of the way in which oppression creates a new oppression. The way out of the crisis is self-determination: for women, for men, for communities, for societies. Only a society which respects the identity of the community can understand and safeguard the autonomy of the women.

7.5.3 Survival of the fittest

Our dominant culture forces feelings of hopelessness and fear on us. Conversations on sustainable development often end in discussions on how to "restrain" ourselves. "People destroy the environment, that's just the way it is", is one common remark. Others say: "The problem is there are just too many people", or "The only thing we can hope for is a big disaster which will teach us to change". And yet others believe that "development is like the struggle for existence: one man's meat is another man's poison". To quote Susan Griffin once more:

> "'All organic beings are exposed to severe competition', it is written. And it is observed that all creatures are pressed into a struggle for existence.
> "That all plants of a given country are at war with one another.
> "That it is the tendency of all beings to multiply faster than their source of nourishment.
> "(Indeed, it is written that the human race tends to outrun subsistence and is kept in bounds only by famine, pestilence or war.)
> "And this struggle is called a natural government, and this warfare is said to lead to perfection.
> "(And it is suggested that war serves 'for the real health of humanity and the building of strong races.')

> "(And it is declared that the history of human society is the history of class struggle. That the collisions between the classes will end in the victory of the proletariat.)
> "(And the development of large corporations, it is pointed out, is also merely the survival of the fittest, merely the working out of a law of nature and a law of God.)" [33]

Susan Griffin's criticism is directed toward the dominant concept of development. That concept is based on competition. As Pronk argues in *A World of Difference:*

> "There are signs of a division, partly splitting the South, the East and, to a lesser extent, the West, into 'front runners', 'hangers-on', and 'stragglers'... 'Stragglers' are the countries whose economies do not have sufficient vitality of their own to stay the pace." (Pronk, p. 61)

The critiques by the women cited above and by the people supporting the Gaia Hypothesis converge in a concept of development based on the restoration of internal control, stability and peaceful co-existence. This concept allows for few external influences: stocks (physical, intellectual, ecological) will be maximized and the movement and export of raw materials (in the form of goods, natural resources and capital) reduced to a minimum. Internal control will consist of a finely intermeshed system of feedbacks. Autonomy, which finds its theoretical roots here in the radical feminist approach, can be translated to the level of communities which determine most of their own flow of life, finding stability on the basis of their own feedback mechanisms, being able to resist being colonized by dominant powers.

7.5.4 Feedback

When a tree is old it will fall down. The forest is ripped open by the crash, light falls on the ground, roots stick into the air. Rain can beat down on the soil once more. The tree rots away and animals, fungi and higher plant varieties live on it. "One man's meat is another man's poison"; there is nothing romantic about it. The death of a tree means the survival of the forest. The large number of species that live in the dying wood ensure diversity; one species will never dominate. There are thousands of species living together, in balance with each other. Just imagine what would happen if the trees did not die. Their foliage would gradually shut out all the sunlight and there would be no undergrowth. There would only be a small number of forest giants. This ecosystem would be very vulnerable. One smart insect could fell all the aged trees and the forest would be destroyed in one blow. Nature's feedback has no purpose, but it is essential.

People have broken the feedback. We chop down the tree, drag it out of the forest for wood or paper production and do nothing to replace it. The forest becomes poorer, the soil exhausted and the number of plant and animal species reduced. Monocultures are extremely sensitive to insects and fungi. If the dead wood (food, in other words) is removed, the soil becomes acidic and the trees become less resistant to pests and diseases. As long as we continue removing the dead trees, the forest as a whole is caught in a downward spiral. People are now introducing an artificial feedback: we throw lime on the forest to raise the pH level, and fertilizer to make it grow. Insecticides are applied to counter the bugs. The forest would be able to restore itself if we only would wait: the only consequence would be that we could no longer remove trees. We are so arrogant as to believe that we can reinvent those complicated feedbacks ourselves. Time and again we prove ourselves wrong: insects and fungi become resistant; the lime we spread does not last long and it disturbs plant and animal life even more. More and more nutrients disappear as a result of the chemical fertilizers which destroy soil structure. In spite of our artificial feedback, the forest dies.

Social feedback mechanisms

The women in Sunita's village felt lonely and powerless. They became isolated because they had been taken out of their traditional communities and were suddenly faced with the extreme violence of a new and hostile society. The violence which occured was a reflection of the force which had crushed their communities. There are people who think that aggression is part of our "being". Maybe that's true, but societies have learned, through tradition, to deal with aggression. Sunita's traditional community had many ways of solving conflicts. The women talked with each other while occupied with their heavy labour on their land. No man was allowed there, so they could exchange essential information freely. Traditionally, men married into the woman's family. In a society in which the men were "officially" the boss, the security of the women was better guaranteed that way. Every community or group of communities has hundreds of these small and large social and economic feedback mechanisms, creating a stable and safe society. The destruction of these feedback mechanisms is responsible for most of the violence in communities all over the world.

7.6 Conflict resolution

Our present complex society, riddled with violence, oppression and a variety of conflicts, is where we have to begin the realization of the main principles of sustainable development.*

The process of achieving sustainable development is complex. In general, this complexity has been underestimated. Again and again, development workers barge in with ready-made ideas, and prance

---

*

1. The ecological principle: development must restore and adapt itself to the diversity of nature, and should be based on the sustainable use of resources.
2. The principle of "cultural and social integrity": development must take place on the basis of internal growth and not be imposed externally.
3. The principle of non-violence: development must be peaceful in the sense that it is not based on direct or structural violence.
4. The principle of emancipation: development must lead to self-reliance, local control over resources and power and participation of marginalized peoples. Similarly it must provide opportunities for carrying out a variety of social and economic activities.
5. The principle of solidarity: development must fulfil the basic needs of life and create liveable circumstances for everyone, it must stimulate equality and prevent asymmetric exchange.
6. The principle of error-friendliness: development must allow for mistakes without resulting in the loss of integrity of resources and the ecosystem.

about with the grace of an elephant in a flower bed. An open approach is needed: it creates much better conditions for growing towards sustainable development than a system of narrow, pre-formulated goals. Projects should not have a tight schedule, starting and finishing at specified times. They should allow plenty of room for local initiatives, and be focused on solving the conflicts. If development workers have a role to play at all, it is to "listen" and to "negotiate".

The cowherds, shepherds and farmers of Gujarat

It hasn't rained in three years. The Vora family lives in one of the suburbs of Rajkot, in Gujarat, North-West India. There is a drought; each family is allowed only one bucket of water a day from the well. The water is muddy and teeming with water fleas and other insect life. In former times Gujarat used to have green pastures as far as the eye could see, and large herds of a local breed of cow, well adapted to the climate and able to provide plenty of milk. Now barren fields stretch towards the horizon – something has gone miserably wrong in the development of Gujarat.

In a small village, I am given lentil soup with chapati, a kind of wheat pancake. The whole village is standing around, watching me eat. A cloud of flies buzz about my head and my intestines are protesting that what I'm eating does not seem safe. The temperature is about 30 degrees centigrade and the air is shimmering.

Most of the people in this village are farmers. They cultivate peanuts for exports. There are a few shepherds and cowherds as well. Formerly, most of the people in the area were shepherds or cowherds, and only the most fertile pieces of land were used to cultivate food for home consumption. Farmers now control most of the land. They spend their earnings on imported food and luxury products. The shepherds and cowherds have access only to the most marginal lands, many of them strips along the sides of the roads. These pasture areas are too small for the number of animals they have to support, and are completely denuded. Even the cacti bordering the peanut fields have been affected. The shepherds and cowherds are confronted by many prejudices: "They are tramps, unwilling to work for their money!"; "They are thieves who don't belong in the community!"

Prior to the privatization of land, and the land reforms which provided farmers with large fields, shepherds and cowherds were very important in Gujarat. They grazed their animals on common lands which were appor-

tioned and used according to local traditions and laws. They have now been relegated to the most marginal lands, and deforestation and desertification are the result.
Farmers who had the means to invest focused on production for export and "chemical agriculture". It was the heyday of the Green Revolution: chemical fertilizers and pesticides were subsidized, peanut farming thrived, and the money came pouring in. As the number of peanut fields increased, desertification became alarming, but the farmers took no notice of this at first. Now, after three years of drought, they are beginning to change their minds. They realize that agricultural practices which undermine the ecological base have no future. They also see that the shepherds and cowherds have no alternative to grazing their sheep and cattle on the poorest lands.

A Gujarati environmental organization is trying to bring the various parties together. The goal is for farmers and cowherds to find a solution to the crisis, but they first have to settle their disputes. A next step is to look for alternatives, where possible at the village level. Youngsters are also drawn into the process. They ask their parents and grandparents to explain once again how the land was used in the past. Young trees are planted by farmers and cowherds, in places where goats cannot damage them. Original tree varieties are used to restore the ties with the past. Rajiv Vora, one of the initiators of the discussion, explains: "It is a long-term project which depends heavily on the participation of both farmers and cowherds. In solving the conflicts we shall have to find new economic alternatives. Our environmental project is primarily a social project – we are sure we will find solutions on the basis of village harmony."

7.7 Looking for a way out

We sometimes get the feeling that only a disaster can end the madness in our way of life. At the same time we realize that the disaster is in fact already taking place, but that we, with our privileges and money, are able to transfer its effects onto people with fewer rights and privileges.

Once in a while I have a discussion with policy makers. They talk about "big money" for the environment, and about creating an "Environmental Facility". The bulk of the money now coming in from environmental taxes will go to this fund, and once it really gets going – once the fund contains hundreds of millions, no, billions of dollars (!) – things will change. Then we will finally have the money to invest in clean technology. It is a tempting thought ... the money ... we have ... can ... if it is properly used ... Wouldn't it be wonderful if this was possible, if this were true? Moments later a colleague defuses the idea and we are ashamed that we believed in it at all.

I sometimes meet people who are genuinely trying to create alternatives. I inhale the dust of the field and listen to their account of the confrontation with hard reality – and later commit it to paper. This reality must be the basis of policy. The stories of men and women farmers, of Rajiv in India, Arnaldo in Argentina, Sumira in Indonesia or Mercedes in Bolivia, all tell of things which have little to do with big money, development aid or the desire for an instructive disaster. Take, for example, reforestation.

Deforestation in the world is immense and ghastly. The statistics of felled trees, millions of tons of fertile topsoil eroded away by wind and rain, and desertification are frightening and make us cry out: do something! The Northerner has a simple solution: "give money and make sure those people plant trees!". Of course it is important for trees to be given another chance. At the same time it does not make much sense to plant trees if the source of deforestation is not tackled. I have seen forest reserves in India which consisted of bare land covered with small hollows. Cowherds had sent their cattle in and the area was denuded. The cowherds were not stupid, thoughtless or aggressive, there was just no other grass. They had reached the brink of survival. In addition, the young seedlings grew on lands they used for grazing. But once a village community is willing to face the problems, solutions are near at hand. A little support may be needed to get things going: small loans or gifts, some research, a publication or a slide show. In some countries the army is deployed to protect the forest against "encroachers". Luckily, the realization is dawning that restoration of the forest starts with the trees planted by villagers and by their resolution to prevent grazing on these lands. If this is the basis, only a few trees will have to be planted, as the soil in many cases contains enough seeds which are ready to take root. Restoration of the forests begins with the people. If the village community succeeds in watching and protecting these woodlands it will have a fully-grown forest within a few decades.

A natural forest provides "services" which a planted forest does not have. The original vegetation contains herbs for medicinal and religious use. Trees are resistant to drought and form a suitable habitat for many insects and other animals. They preserve the soil, ensure a more constant water supply and regulate the climate. A village which decides to maintain a forest has gained a great deal in terms of social organization. The forest is not the main objective: it is far more important to restore the special relationships between human beings, plants and animals. This is a very complicated process, and the various forces involved can reinforce each other in a positive or in an utterly negative sense (positive and negative synergy).

Dam in Rajasthan, India photo: Rozenburg/HH

This example illustrates the importance of the relationship between people and their surroundings. The purpose of change is based on the restoration of this relationship. Big money, however well meant, cannot revive it – it may well result in uncoupling the relationship even further. Donors don't enjoy hearing this, especially when environmental issues are fashionable and a lot of money is not only available, but has to be spent. "Donor-driven development" has produced a whole series of failures.

This does not mean financial transfers should be totally taboo. We should realize, however, that transfers are not neutral and always result in some parties receiving more power and others less. Whenever money is really needed, for the modernization of a city's electricity grid for example, it is important to recognize that this will give rise to inequality. It is just too easy to pump hundreds of millions into so-called "environmental projects" while we know very well that a few leading industrialists will run away with the money to develop their own enterprise (to the detriment of others).

Development cooperation is concerned with more than just the transfer of funds. The prime issues are trust in the integrity of the relationship between people and the environment, and an understanding of the factors which contribute to the disturbance of the ties. Westerners have learned to distrust and even exploit the environment, and consider the cultures which stress the closeness of these ties to be primitive. An Indian friend recently told me the following story. A wise Chinese philosopher dreamed he had become a butterfly. "Maybe I died and have been reincarnated in the butterfly I see flying over there", he said the next day. "But now I am awake I can imagine that the butterfly may be thinking it is reincarnated in me." The organic relationship between all creatures is expressed in the idea of reincarnation, which is found in many religions. It demonstrates a deep feeling of relationship with the surrounding world – a relationship Western people will have to regain if the world is to change fundamentally.

Changing our future

People will remain constrained individuals until people join together and organize themselves. As an organized group they can build pyramids, change the flow of ocean currents and disrupt the climate. This is the power we need to be able to create change and move towards sustainability, in the North and in the South. The change is not going to take place on the basis of a moral understanding of the way the world works; it will take place on the basis of an immediate involvement in the struggle to survive. Not the fittest, but those who fit, will survive. Creativity for change will not emerge from the fear of being destroyed or from a violent and enforced pressure to change. It will flow from the practical changes being developed by communities. This is one of the fundamental principles of sustainable development.

Within the framework described in this book, there are three different approaches to society. One is the open market approach, in which market forces determine the course of events. The second approach is the strong state market, in which decisions are made by a central organization and the centre bears the responsibility for change. The third approach is a society of responsible communities, and the centre of change occurs at the local community level. This third approach resembles autonomy, or de-linking, as I described it. It is a people-oriented approach, which is the backbone of many grass-roots movements.

> I live in a community, in the centre of Haarlem, in the Netherlands. We are concerned about the environment, and because we realize our own consumption is responsible for a great deal of destruction, we are trying to live differently. Forty people live in this community.
> When we visited our greengrocer to ask if he could provide us with unsprayed potatoes, which he didn't stock, he found unsprayed potatoes for us. It wasn't just his desire to win customers that sent him looking for biologically grown potatoes, he too shared our concern. A few weeks later he proudly announced that he would also provide us with biologically grown onions and biologically grown carrots. When I recounted this story at a meeting of an alliance of local solidarity groups, people came to the realization that my household was able to do this because we were a group. Individuals could not have the same effect on their greengrocers. In the course of the discussion we realized that every neighbourhood had its own community centre and community organizations. If we could spend as much effort in reaching these groups as we did in setting up political lobbies at a

national and an international level, we could have an enormous impact! The material produced by our national movements was an inadequate tool in creating a link with our own communities. That material contained information on endless discussions which were interesting for people who worked at the level of the United Nations, and were concerned about bio-diversity, climate change and the hole in the ozone layer. What we needed was material urging communities to buy locally, to buy responsibly, to discuss biological gardening, biological cooking, using less-polluting types of paint, other means of transport. We needed to create a market in our city for products we think are worth selling. We needed to practise being the other society we so fervently want to create. This, we decided, was the best approach to campaigning in the context of the UNCED meeting.

As I was cycling to my work after that meeting, I realized that this is what Ricardo from Lampa in Chile was doing. Wasn't he a member of an initiative of Ecological Communities in Chile? Why didn't we realize much earlier that it is not only in the South that community projects are sorely needed? Why didn't we recognize the importance of our own communities, our identity in the North? Why are we so slow in learning from the grass-roots experience around the world?

The process of political lobbying, of spending creativity in changing the opinions of national and world political leaders, can be important. But these efforts should in the final event bring about the realization that communities have to change from within. The Paris NGO conference on Environment and Development, part of the preparatory process toward UNCED, adopted a resolution stating:

> "Citizens' groups, community groups, environment, development and consumers organizations across the world – as distinct from organizations associated with profit-making institutions – are united in their single most important desire of expanding the role of civil society in governance, and reinhabiting our life places in a socially just and ecologically sustainable manner. We must work together even more intensively than we have done in the past, both within our regions and across national boundaries, to change the behaviour of national élites and state, international and commercial organizations. But it is important that we further democratize our own work patterns. Urban and issue-based groups must work closely with community groups and/or constituency-based citizen's groups, and carefully avoid adopting a representative role without an appropriate process of legitimization. Similarly, Northern citizens' groups, however well

> intentioned, must work closely with, and consult with, Southern and Eastern citizens' groups before taking up issues of global concern on their behalf. Relations between citizens' groups from the North and South should be built on true and equal partnership and they should not mimic the dominance-dependence relationship of the governments of the North and South. There is a lot we can learn from each other and it is our bounden duty to document our struggles and our successes and failures and keep each other informed."

This informed and creative grass-roots movement can be the basis from which we discuss the role of price mechanisms, and the role of the state in regulating emissions and enforcing standards. It is a top priority to strengthen these autonomous organizations, in the North, in the South and in the East.

Notes

1. I used extensively the book *Mothers for Life — Motherhood and marginalization in the North Central Province of Sri Lanka* by Joke Schrijvers, Delft, The Netherlands, 1986.
2. World Commission on Environment and Development, *Our Common Future,* Oxford Press, UK, 1987.
3. Tjeerd Deelstra (ed), Jack Overkamp and Ton Koning, 'Environmental Aspects of Citification in Development Cooperation', an exploratory study commissioned by the International Institute for urban Development, Delft, The Netherlands, June 1989.
4. Tjeerd Deelstra *et al* 'Human Settlements and Sustainable Development in the Third World', Directorate General for Housing, The Hague, The Netherlands, 1990.
5. Vereniging Milieudefensie and Stichting Natuur en Milieu, 'Het broeikasteffekt, erop of eronder', 1990.
6. Ministry of Housing, Physical Planning and Environment, 'Climate Change Policy in the Netherlands and Supporting Measures', Leidschendam, November 1990.
7. Kalverkamp and van Hoytema, 'In search of sustainable agriculture, an outline of profits and losses of conversion to biodynamic agriculture in the Netherlands, November 1990.
8. Many thanks to Ian Tellam of the World Information Service on Environment and Development in Amsterdam, for material he has collected.
9. Lord Lever *et al,* 'The debt crisis and the World Economy', Commonwealth Secretariat, London, 1984.
10. Morgan Guaranty, cited in Denham, J., 'Capital Flight', European Appeal Briefing Paper No. 2, Fondad, 1989.
11. UN Department of Public Information, *The Debt Crisis,* 1989.
12. Clark, J. and Keen, D., 'Debt and Poverty: a case study in Zambia', Oxfam, 1988.
13. Denham, J., 'Out of their debts: UK banks and the Third World debt crisis', War on Want Campaigns, London, 1989.
14. Jackson, B., *Poverty and the Planet,* Penguin, London, 1990.

15. Madden, P., *Debt and Deforestation,* Christian Aid Research Unit, UK, August 1990.
16. Hecht, S. and Cockburn, A., *The Fate of the Forest,* Penguin, London, 1991.
17. George, S., *A Fate Worse than Debt,* Pelican, London, 1988.
18. Leonard Frank, 'The Development Game', Granta, date unknown to author.
19. Wolfgang Sachs, *On the Archaeology of the Development Idea,* November 1989, The Pennsylvania State University, USA, pp. 1-2.
20. Corinne Kumar-D'souza, 'The South Wind', *Sangarash* No. 3, 1989.
21. Ted Trainer, *Developed to Death, Rethinking Third World Development,* Green Print, London, 1989.
22. Sachs, 'On Archaeology' *op cit.*
23. Herman E. Daly and John B. Cobb, Jr., *For the Common Good, redirecting the economy toward community, the environment and a sustainable future,* Beacon Press, Boston, 1989/Green Print, London, 1990.
24. Daly and Cobb, *For the Common Good, op cit,* pp. 48-49.
25. Daly and Cobb, *ibid.*
26. Wolfgang Sachs, 'The Gospel of Global Efficiency, On Worldwatch and other Reports on the State of the World', in *Transnational Associations* 3/1989.
27. Karl Polanyi *The Great Transformation,* (1944, reprint Beacon Press, Boston, 1957): quoted in Daly and Cobb, *ibid.*
28. Sachs, 'The Gospel of Global Efficiency', quoted in Daly and Cobb, *ibid.*
29. John Maynard Keynes, 'National Self Sufficiency', 1933. In: *The Collected Writings of John Maynard Keynes,* Vol. 21. Edited by Donald Moggeridge. London: Macmillan Cambridge University Press, for the Royal Economic Society pp. 235-236.
30. Trainer, *Developed to Death, op cit.*
31. J. Falconer, 'The major significance of minor forest products: local people's use and values of forest in the West African Humid Forest Zone', Forestry Department, FAO, Rome. Quoted in De Beer and McDermott, *The Economic Value of Non-Timber Forest Products in Southeast Asia,* 1989.
32. Susan Griffin, *Woman and Nature, the Roaring Inside Her,* The Women's Press, 1984.
33. Susan Griffin, *op cit.*
34. See footnote 1.